# Logic, Argumentation & Reasoning

## Interdisciplinary Perspectives from the Humanities and Social Sciences

## Volume 26

Logic, Argumentation & Reasoning (LAR) explores links between the Humanities and Social Sciences, with theories (including decision and action theory) drawn from the cognitive sciences, economics, sociology, law, logic, and the philosophy of science.

Its main ambitions are to develop a theoretical framework that will encourage and enable interaction between disciplines, and to integrate the Humanities and Social Sciences around their main contributions to public life, using informed debate, lucid decision-making, and action based on reflection.

- Argumentation models and studies
- Communication, language and techniques of argumentation
- Reception of arguments, persuasion and the impact of power
- Diachronic transformations of argumentative practices

LAR is developed in partnership with the Maison Européenne des Sciences de l'Homme et de la Société (MESHS) at Nord - Pas de Calais and the UMR-STL: 8163 (CNRS).

This book series is indexed in SCOPUS.

Proposals should include :

- A short synopsis of the work, or the introduction chapter
- The proposed Table of Contents
- The CV of the lead author(s)
- If available: one sample chapter

We aim to make a first decision within 1 month of submission. In case of a positive first decision, the work will be provisionally contractedŮthe final decision about publication will depend upon the result of an anonymous peer review of the complete manuscript.

The complete work is usually peer-reviewed within 3 months of submission.

LAR discourages the submission of manuscripts containing reprints of previously published material, and/or manuscripts that are less than 150 pages / 85,000 words.

For inquiries and proposal submissions, authors may contact the editor-in-chief, Shahid Rahman at: shahid.rahman@univ-lille.fr, or the managing editor, Juan Redmond, at: juan.redmond@uv.cl

Lorenzo Magnani

# Discoverability

## The Urgent Need of an Ecology of Human Creativity

Springer

Lorenzo Magnani
Dipartimento di Studi Umanistici
University of Pavia
Pavia, Italy

ISSN 2214-9120 ISSN 2214-9139 (electronic)
Logic, Argumentation & Reasoning
ISBN 978-3-030-93328-9 ISBN 978-3-030-93329-6 (eBook)
https://doi.org/10.1007/978-3-030-93329-6



This Springer imprint is published by the registered company Springer Nature Switzerland AG
The registered company address is: Gewerbestrasse 11, 6330 Cham, Switzerland

*Our faculty of guessing corresponds to a bird's musical and aeronautic powers; that is, it is to us, as those are to them, the loftiest of our merely instinctive powers. I suppose that if one were sure of being able to discriminate between the intimations of this instinct and the self-flatteries of personal desire, one would always trust to the former. For I should not rate high either the wisdom or the courage of a fledgling bird, if, when the proper time had come, the little agnostic should hesitate long to take his leap from the nest on account of doubts about the theory of aerodynamics.*

*Charles Sanders Peirce, Collected Papers, 7.48, 1907*

*To my daughter Lorenza and to my grandson Valentino*

# Preface

We academics all know that philosophers of science are not involved—neither directly nor indirectly—in decisions regarding public policy: it seems that policy decisions must be based on social goals and other factors that have nothing to do with epistemology. At this point, it seems that various aspects of current decision making in commercialized scientific research, for example, in biomedical area, are jeopardizing discoverability in science and therefore scientific creativity and human creative abduction in general. Furthermore, what we can call the "perfect epistemic storm"—which is not directly addressed in this book and has been generated by the many failures and confusions happened during this corona virus era—is visible to everyone. As an epistemologist, I am just indicating the problem of protecting discoverability, which I also think is at the base of the epistemic disasters of the present time of the pandemic: too many aspects of the socio-economical organization of science—increasingly business oriented—put in danger the genuine tradition of objective science and some of the best aspects of the human abductive scientific cognition. I do not have "practical" suggestions and solutions, but surely the smart people who are involved in economic and political decisions and at the same time are interested in favoring the further growth of science and civilization in the Western world will find ones.

The book addresses the problem of discoverability and consequently of the sustainability of human creativity in an "eco-cognitive" perspective, with special attention to science, and concentrates on the question of hypothesis generation and choice, that is, on creative abduction and on the conditions of its possibility, traditionally disregarded by philosophy, logic, epistemology, and cognitive science. I think the intellectual, didactic, but also "social" virtues of providing an analysis of discoverability are extremely important: having a better knowledge of discoverability—for example, in science—enables a better promotion of human creativity, which is the condition of possibility of increasing other kinds of knowledge, always needed in our technological era. Reaching higher degrees of discoverability should also shed new light on the current negative challenges to human creativity and consequently on the obstacles to the cognitive and social flourishing of human beings.

To further illustrate the importance of sustaining human creativity through the enhancement of discoverability, the example of the recent Encyclical Letter of the Holy Father Francis, head of the Catholic Church, is very useful. In the recent Encyclical Letter, Laudato Si' of the Holy Father Francis on Care for Our Common Home the Pope quotes several times the importance of *human creativity* in solving the many problems of humanity that are illustrated. Creativity is seen as a fundamental tool necessary to fix the failures of our societies and of our lives. Just to make some examples, we find the following sentences: "In order to continue providing employment, it is imperative to promote an economy which favours productive diversity and business creativity" (129), "creativity should be shown in integrating rundown neighbourhoods into a welcoming city" (152), "[...] political and institutional frameworks do not exist simply to avoid bad practice, but also to promote best practice, to stimulate creativity in seeking new solutions and to encourage individual or group initiatives" (177). Moreover, the awareness concerns the fact that in current societies, creativity is jeopardized, and this is exhibited: speaking of the effect of technology, we find, for example, the following sentences "Our capacity to make decisions, a more genuine freedom and the space for each one's alternative creativity are diminished" (108) "Human creativity cannot be suppressed" (131).

Chapter four of the Encyclical Letter introduces the need for an *integral ecology*, also necessary in education, and many subtypes of ecology are illustrated: first of all, environmental, economical, and social ecology, but also cultural ecology, ecology of daily life, human ecology, and ecology of man ("man too has a nature that he must respect and that he cannot manipulate at will"—155). In all these endeavors, related to the emancipation of human lives, human creativity is so fundamental that I would strongly suggest to *integrate* the Encyclical deontological commitments with what I call *ecology of human creativity*.

My proposal of an integration of the Encyclical deontological commitments actually resorts to the need of accepting that an ecology of human creativity has priority over the others illustrated in the Encyclical Letter: consequently, a first ecological duty is the one that protects and sustains human creativity because it is exactly human creativity that can grant the implementation of the other kinds of ecology I have just quoted from the Encyclical letter. I consider the *ecology of human creativity* a *conditio sine qua non* in the following sense: I am convinced that without creativity and skillful human capacities, all the other envisaged and invoked ecologies and sustainability, in general, tend to sadly fail. Sustaining the eco-cognitive conditions of discoverability is fundamental because only creativity can grant the implementation of the other kinds of ecology, which are urgently at play in our era, so that the ecological improvement can in turn positively affect human creativity.

In the present book, I will, for example, illustrate the importance of sustaining the so-called epistemic niches, that is, those eco-settings that have the potential for scientific creativity. I will center my attention on scientific creativity because of its importance in our current societies and collectives, not only because of its intrinsic—so to speak—value but also because in our technological era, full of science-based artifacts of various type, knowledge in general, and scientific knowledge in particular are fundamental. In my book *Morality in a Technological World. Knowledge as Duty* (Magnani, 2007), I treated this issue extensively, arguing that the maintenance and flourishing of technological societies require a great deal of new scientific and ethical knowledge as well as modern approaches to moral deliberation, and to achieve these goals I provided compelling analyses, also offering a variety of strategies we might use to solve them. I believe that producing and applying recalibrated appropriate general, scientific, and moral knowledge has become a duty. It is in this vein that I analyzed troubling issues such as cyberprivacy, globalization, bad faith, cloning, biotechnologies, and ecological imbalances: the right creative knowledge, I think, can manage these challenges and counter many of technology's ill effects by preserving ownership of our own destinies, encouraging responsibility, and enhancing freedom. However, creative knowledge can completely flourish only when discoverability is sustained, protected, and enhanced.

A consequence of the above considerations is the following: if creativity is jeopardized, and in particular scientific creativity, it is practically impossible to acceptably and successfully deal with our technological societies, also from an ethical perspective: this means that the need of an *ecology of human creativity* I am endorsing here is rooted in the observations I have just sketched, quoting my book of 2007.

The present book, which aims at improving knowledge of discoverability, will take into account:

- the *distributed* and *embodied* nature of scientific cognition, ultimately related to the idea of the importance of the external cognitive tools and mediators in cognition;
- the central role of the dynamics of the production and of the rational handling of hypotheses, by referring to the various multimodal aspects of *abduction*, visual/diagrammatic, verbal-propositional, emotional, and manipulative;
- the fact that science is characterized by a *maximization of abducibility*, performed thanks to specific constraints.

These topics are analyzed in terms of what I consider the main tenets of an eco-cognitive approach to the epistemology of discoverability:[1]

1. *Chapter one*
   Fruitfully approaching the problems of discoverability involves an important intermediate step, which concerns the role of abductive cognition, that is, reasoning to hypotheses, and the logical models of it. To this aim, when engaged in formalizing abductive reasoning, it is extremely useful to see inferences adopting the more general concepts of input and output instead of those of premisses and conclusions. Indeed, from this perspective, abductive inferences can be first of

[1] A considerable part of the recent academic literature—for example in social epistemology—refers the word epistemology to the whole area of cognitive reasoned activities. In this book, I basically adopt its classical intended meaning, which refers only to scientific cognition.

all seen as related to logical processes in which input and output fail to hold each other in an expected relation, with the solution involving the modification of inputs, not that of outputs. Unfortunately, if input and output fail to hold each other in a "good" relation, it becomes very difficult to solve the related abductive problem: discoverability is jeopardized. In this perspective—and given the fact science produces and "maximizes" cognition through a process in which affirming truths implies negating truths—the analysis of abductive processes leads us to the emphasis on the importance of the following main aspects: "optimization of eco-cognitive situatedness," "maximization of changeability" of both input and output, and high "information-sensitiveness." I will also illustrate that irrelevance and implausibility are not always offensive to reason, and so they can favor both discoverability and discovery. For example, we cannot be sure, more broadly, that our guessed hypotheses are plausible (even if we know that looking—in advance—for plausibility is a human good and wise heuristic), indeed an implausible hypothesis can, later on, result plausible.

2. *Chapter two*
   Recent research in the area of the so-called EEEE cognition (extended, embodied, embedded, enacted) has shown that human cognition and its evolutionary dimension can be better understood in terms of their environmental situatedness. This means humans do not retain in their memory rich representations of the environment and its variables, but they actively manipulate it by picking up information and resources upon occasion, already available, or extracted/created and made available: information and resources are not only given, but they are actively sought and even manufactured. In this sense, we can consider human cognition as a chance-seeker system. Consequently, in my terminology, chances are not simply information, but they are also "affordances", namely environmental anchors that allow us to better exploit external resources. Discoverability is of course related to the availability of the appropriate affordances. Abduction is still important: it describes all those human and animal hypothetical inferences that are operated through actions made up of smart manipulations of the environment to both detect new affordances and to create manufactured external objects that offer new affordances/cues. A final section deals with the defense of the inferential view of abduction, affordances, and cognitive niche construction (based on epistemology and cognitive science) and their importance for the analysis of discoverability, against the reductionist attitudes due to some interpretations provided by the predictive processing theory, based on the idea of a Bayesian brain.
3. *Chapter three*
   In this chapter, with the help of the concepts of locked and unlocked strategies, abduction, and optimization of eco-cognitive openness, I will describe some central aspects of the cognitive character of reasoning strategies and related heuristics, with the aim of shedding new light on the cognitive aspects of deep learning machines and their impact on discoverability, and so creativity. Taking advantage of my studies on abduction, I will contend that what I call *eco-*

*cognitive openness* is undermined in the case of famous computational programs such as AlphaGo/AlphaZero, because they are based on *locked abductive strategies*. Unlocked abductive strategies, which are in tune with what eco-cognitive openness requires, qualify those high-level kinds of abductive creative reasoning that are typical of human-based cognition, and locked abductive reasoning strategies are instead much simpler than the unlocked ones to be rendered at the computational level. The second part of the chapter emphasizes, taking advantage of the logical and cognitive studies introduced in the previous chapters, the importance of *knowledge in motion*—in multidisciplinary, interdisciplinary, and transdisciplinary scientific research—and its role in favoring high-level kinds of discoverability, and so of abductive results and of the so-called "optimization of eco-cognitive openness and situadedness." The concept of "knowledge in motion" is also the necessary conceptual premise of the analysis of various kinds of "*epistemic irresponsibility*," which are illustrated in chapter four.

4. *Chapter four*
   In this chapter I will analyze some important aspects of the organization of research and development (R&D) in the case of biopharmaceutical companies, which represent a prototypical situation of what I call impoverished epistemic niches. At least in this case we clearly see a challenge to the epistemic integrity of modern science. Taking advantage of the logical and cognitive studies illustrated in the previous chapters, which emphasize the crucial role played in abductive cognition by the so-called "optimization of eco-cognitive openness and situadedness," this chapter will describe the hot problem of the current emergence of disparate kinds of "*epistemic irresponsibility*" that strongly jeopardize discoverability.[2] Interesting cases related to the commodification and commercialization of science, marketing of technoscientific products, and impoverishment of the so-called epistemic niches are illustrated, which show that human fruitful abductive cognition in science is increasingly assaulted and jeopardized, and at the same time human creativity is seriously endangered.
5. *Chapter five*
   Taking advantage of the logical and cognitive studies illustrated in the previous chapters, which emphasize the crucial role played in abductive cognition by the so-called "optimization of eco-cognitive openness and situadedness," "knowledge in motion," and the concept of "epistemic irresponsibility," the present chapter will introduce the concept of overcomputationalism, to help interpret the related concepts of pancognitivism, paninformationalism, and

[2] In this chapter I prefer to adopt the expression epistemic irresponsibility (instead of epistemological irresponsibility), because I attribute to the adjective epistemic a restricted meaning. "Epistemic" pertains to scientific knowledge or the conditions for acquiring it, which involve eco-cognitive situations in which not only scientists but also other economical, political, and institutional agents directly or indirectly intertwined with scientific research are involved. Instead, "epistemological" also expressly refers to the philosophical community of epistemologists, who in general are not affected by epistemic irresponsibility, but by other problems such as, for example, philosophical irrelevance or scholasticism.

pancomputationalism and their impact on discoverability. In the second part of the chapter, I will submit to the attention of the reader a question that in my opinion synthesizes many of the problems described in this book: will the future of eco-cognitive settings be computationally tailored or humanly tailored? The challenges against human creative abduction and epistemic rigor on the part of what I call computational invasive "subcultures" and unwelcomed effects of selective ignorance are illustrated. Finally, I will describe two fundamental roles played by human knowledge and the need of their continuous enhancement to the aim of sustaining discoverability: to maximize the ownership of their own destinies, thanks to everyday, philosophical, and scientific knowledge about natural and artificial phenomena human beings have improved their capacities to track the external world; in turn thanks to ethical knowledge humans have learnt to track the behavior of other human beings to the aim of favoring cooperation.

I am grateful to many colleagues and friends for their helpful suggestions and much more. For valuable comments and discussions, I am indebted and grateful to John Woods, Paul Thagard, Ping Li, Atocha Aliseda, Woosuk Park, Nancy J. Nersessian, Giuseppe Longo, Yukio Oshawa, Akinori Abe, Michael Leyton, Dov Gabbay, John Josephson, Walter Carnielli, Gerhard Schurz, Balakrishnan Chandrasekaran, Jon Williamson, Douglas Walton, Cameron Shelley, Sami Paavola, Luís Moniz Pereira, Thomas Addis, Giovanna Magnani, Lorenza Magnani, Diderik Batens, Joke Meheus, Simon Colton, Athanassios Raftopoulos, Michael Hoffmann, Ilkka Niniluoto, Theo A. F. Kuipers, Chris Sinha, Ryan D. Tweney, Ferdinand D. Rivera, Peter Flach, Antony Kakas, Oliver Ray, Luis A. Pineda, Atsushi Shimojima, Pat Langley, Demetris P. Portides, and to my collaborators Tommaso Bertolotti, Selene Arfini, and Alger Sans Pinillos. Also of great help was correspondence or conversation with a number of people whose influence on various of the book's particularities is more or less palpable. I especially wish to acknowledge Hanne Andersen, Otávio Bueno, Marco Buzzoni, Sanjay Chandrasekharan, Marcelo Dascal, Gordana Dodig Crnkovic, Mauro Dorato, Michel Ghins, Marcello Guarini, Ricardo Gudwin, Albrecht Heeffer, Mireille Hildebrandt, Shahid Rahman, Kenneth Einar Himma, Fabio Minazzi, Gerhard Minnameier, Margaret Morrison, Alfredo Pereira, Ahti-Veikko Pietarinen, Dagmar Provijn, João Queiroz, Chiaki Sakama, Colin Schmidt, Nora Schwartz, Frederik Stjernfelt, Mauricio Suárez, Jeroen van den Hoven, Peter-Paul Verbeek, Riccardo Viale, and Marion Vorms. The preparation of the volume would not have been possible without the contribution of resources and facilities by the Computational Philosophy Laboratory (Department of Humanities, Philosophy Section, University of Pavia, Italy). This project was conceived as a whole, but as it developed, various parts have become articles, which have now been excerpted, revised, and integrated into the current text.

Parts of this book are excerpted from L. Magnani (2015), The eco-cognitive model of abduction. Απαγωή now: Naturalizing the logic of abduction. *Journal of Applied Logic*, 13, 285–315. L. Magnani (2016), The eco-cognitive model of abduction II. Irrelevance and implausibility exculpated, *Journal of Applied Logic* 15, 94–129; (chapter one); L. Magnani (2019), AlphaGo, locked strategies, eco-cognitive openness, in H. Zenil and S. Bringsjord, eds., "Philosophy and Epistemology of Deep Learning" a special issue of *Philosophies* 2019, 4(1), 8, https://doi.org/10.3390/philosophies4010008 (chapter three); L. Magnani, (2009), *Abductive Cognition. The Epistemological and Eco-Cognitive Dimensions of Hypothetical Reasoning*, Springer (chapter two); L. Magnani, (2017), *The Abductive Structure of Scientific Creativity. An Essay on the Ecology of Cognition*, Springer (chapter four), L. Magnani (2021), Computational domestication of ignorant entities. Unconventional cognitive embodiments, *Synthese* 198, 7503–7532, Special Issue edited by L. Magnani and S. Arfini "Knowing the Unknown: Philosophical Perspectives on Ignorance," (chapter five). I am grateful to Elsevier and Springer for permission to include portions of previously published articles.

Pavia, Italy
July 2021

Lorenzo Magnani

## Reference

Magnani, L. (2007). *Morality in a technological world. Knowledge as duty*. Cambridge: Cambridge University Press.

# Contents

# Chapter 1
# Discoverability Explained

## Optimizing the Eco-Cognitive Situatedness

Fruitfully approaching the problems of discoverability involves an important intermediate step, which concerns the role of abductive cognition, that is reasoning to hypotheses and the logical models of it. To this aim, when engaged in formalizing abductive reasoning, it is extremely useful to see inferences adopting the more general concepts of input and output instead of those of premisses and conclusions, which are standardly used to characterize abduction in the syllogistic scheme of the fallacy of "affirming the consequent". As I will better explain below, we can say that of course the relation from premisses to conclusions still represents a particular case of input-output relation, but seeing abductive inferences in the input/output framework helps to abandon the habit—typical of the image of abductive inference as composed by premisses and a conclusion—of privileging the reaching of an output as the abductive conclusion. On the contrary, adopting the view of inference in terms of input and output can highlight the *process of modifying part of the inputs, or both inputs and outputs*, in order to obtain the desired relation between inputs and outputs. Indeed, from this perspective abductive inferences can be first of all seen as related to logical processes in which input and output fail to hold each other in an expected relation, with the solution involving the modification of inputs, not that of outputs. Unfortunately, if input and output fail to hold each other in a "good" relation, very difficult is to solve the related abductive problem: discoverability is jeopardized. In this perspective—and given the fact science produces and "maximizes" cognition through a process in which affirming truths implies negating truths—the analysis of abductive processes leads us to the emphasis on the importance of the following main aspects: "optimization of eco-cognitive situatedness", "maximization of changeability" of both input and output, and high "information-sensitiveness". I will also illustrate that irrelevance and implausibility are not always offensive to reason and so they can favor both discoverability and discovery. For example, we cannot be sure, more broadly, that our guessed hypotheses are plausible (even if we know that looking—in advance—



L. Magnani, *Discoverability*, Logic, Argumentation & Reasoning 26,
https://doi.org/10.1007/978-3-030-93329-6_1

for plausibility is a human good and wise heuristic), indeed an implausible hypothesis can, later on, result plausible.

## 1.1 Sustaining Discoverability: Towards an Ecology of Human Creative Cognition

To further illustrate the importance of sustaining human creativity through the enhancement of discoverability the example of the recent Encyclical Letter of the Holy Father Francis, head of the Catholic Church, is very useful. In the recent Encyclical Letter Laudato Si' of the Holy Father Francis on Care for Our Common Home the Pope various times quotes the importance of *human creativity* in solving the several problems of humanity that are illustrated. Creativity is seen as a fundamental tool necessary to fix the failures of our societies and our lives. Just to make some examples, we find the following sentences: "In order to continue providing employment, it is imperative to promote an economy which favours productive diversity and business creativity" (129), "creativity should be shown in integrating rundown neighbourhoods into a welcoming city" (152), "[...] political and institutional frameworks do not exist simply to avoid bad practice, but also to promote best practice, to stimulate creativity in seeking new solutions and to encourage individual or group initiatives" (177). Moreover, the awareness concerning the fact that in current societies creativity is jeopardized is exhibited: speaking of the effect of technology we find for example the following sentences "Our capacity to make decisions, more genuine freedom and the space for each one's alternative creativity are diminished" (108) "Human creativity cannot be suppressed" (131).

Chapter four of the Encyclical Letter introduces the need of an *integral ecology*, also necessary in education, many subtypes of ecology are illustrated: first of all environmental, economical, and social ecology, but also cultural ecology, ecology of daily life, human ecology, ecology of man ("man too has a nature that he must respect and that he cannot manipulate at will"—155). In all these endeavors, related to the emancipation of human lives, human creativity is so fundamental that I would strongly suggest to *integrate* the Encyclical deontological commitments with what I call *ecology of human creativity*.

My proposal of integration of the Encyclical deontological commitments resorts to the need of accepting that this kind of ecology of human creativity has priority over the others illustrated in the Encyclical Letter: consequently, a first ecological duty is the one of protecting and sustain human creativity because it is exactly human creativity that can grant the implementation of the other kinds of ecology I have just quoted from the Encyclical letter. I consider the *ecology of human creativity* a *conditio sine qua non* in the following sense: I am convinced that without creativity and skillful human capacities all the other envisaged and invoked ecologies and sustainability, in general, tend to sadly fail.

In the present book, I will for example illustrate the importance of sustaining the so-called epistemic niches, that is of those eco-settings that are the condition of possibility of scientific creativity. I will center my attention on scientific creativity because of the importance of it in our current societies and collectives, not only because of its intrinsic—so to speak—value, but also because, in our technological era, full of science-based artifacts of various types, knowledge in general and scientific knowledge in particular are fundamental. In my book *Morality in a Technological World. Knowledge as Duty* (Magnani, 2007) I extendedly treated this issue, arguing that the maintenance and flourishing of technological societies requires a great deal of new scientific and ethical knowledge as well as modern approaches to moral deliberation, and to achieve these goals I provided compelling analyses also offering a variety of strategies we might use to solve them. I believe that producing and applying recalibrated appropriate general, scientific, and moral knowledge has become a duty. It is in this vein that I analyzed troubling issues such as cyberprivacy, globalization, bad faith, cloning, biotechnologies, and ecological imbalances: the right creative knowledge, I think, can manage these challenges and counter many of technology's ill effects by preserving ownership of our own destinies, encouraging responsibility, and enhancing freedom. However, creative knowledge can completely flourish only when discoverability is sustained, protected, and enhanced.

A consequence of the above considerations is the following: if creativity is jeopardized and in particular scientific creativity, it is practically impossible to acceptably and successfully deal with our technological societies, also from an ethical perspective: this means that the need of an *ecology of human creativity* I am endorsing here is rooted in the observations I have just sketched, quoting my book of 2007.

## 1.2 The Eco-Cognitive Model of Abduction (EC-model)

At the center of my perspective on cognition and consequently of abduction is the emphasis on the "practical agent", on the individual agent operating "on the ground", that is, in the circumstances of real life. In all its contexts, from the most abstractly logical and mathematical to the most roughly empirical, I always emphasize the cognitive nature of abduction. Reasoning is something performed by cognitive systems. At a certain level of abstraction and as a first approximation, a cognitive system is a triple $(A, T, R)$, in which $A$ is an *agent*, $T$ is a *cognitive target* of the agent, and $R$ relates to the *cognitive resources* on which the agent can count in the course of trying to meet the target-information, time and computational capacity, to name the three most important. My agents are also *embodied distributed cognitive systems*: cognition is embodied and the interactions between brains, bodies, and external environment are its central aspects. Cognition is occurring taking advantage of a constant exchange of information in a complex distributed system that crosses the boundary between humans, artifacts, and the surrounding environment, where also instinctual and unconscious abilities play an important role. This interplay is especially manifest and clear in various aspects of abductive cognition, that is in reasoning to hypotheses.

My perspective adopts the wide Peircean philosophical framework, which approaches "inference" *semiotically* (and not simply "*logically*"): Peirce distinctly says that all inference is a form of sign activity, where the word sign includes "feeling, image, conception, and other representation" (Peirce, 1866–1913, 5.283). It is clear that this semiotic view is considerably compatible with my description of cognitive systems as embodied and distributed systems. It is in this perspective that we can fully appreciate the role of abductive cognition, which not only refers to propositional aspects but it is also performed in a framework of distributed cognition, in which also models, artifacts, internal and external representations, and manipulations play an important role. Already in a passage concerning abduction (that is ἀπαγωγή—"leading away") of chapter B25 of the Aristotelian *Prior Analytics*, we can see some of the current well-known distinctive characters of abductive cognition, which are in tune with my eco-cognitive model of abduction (EC-Model), I will soon describe. Aristotle is already pointing to the fundamental inferential role in reasoning of those externalities that substantiate the process of "leading away" (ἀπαγωγή) . We can gain a new positive perspective about the "constitutive" eco-cognitive character of abduction, just thanks to Aristotle himself.[1]

The backbone of this approach can be found in the manifesto of my Eco-Cognitive Model (EC-Model) of abduction in (Magnani, 2009).[2] It might seem awkward to speak of "abduction of a hypothesis in literature," but one of the fascinating aspects of abduction is that not only it can warrant for scientific discovery, but for other kinds of creativity as well. We must not necessarily see abduction as a *problem-solving device* that sets off in response to a cognitive irritation/doubt: conversely, it could be supposed that esthetic abductions (referring to creativity in art, literature, music, games, etc.) arise in response to some kind of esthetic irritation that the author (sometimes a *genius*) perceives in herself or in the public. Furthermore, not only esthetic abductions are free from empirical constraints in order to become the "best" choice: many forms of abductive hypotheses in traditionally-perceived-as-rational domains (such as the setting of initial conditions, or axioms, in physics or mathematics) are relatively free from the need of an empirical assessment. The same could be said of moral judgments: they are eco-cognitive abductions, inferred upon a range of internal and external cues and, as soon as the judgment hypothesis has been abduced, it immediately "can" become prescriptive and "true," informing the agent's behavior as such.

The kinds of abduction that do not need empirical confirmation can be called "knowledge enhancing": Peirce implicitly provides various justifications of the

[1] I have illustrated in detail the various aspects, philosophical, logical, and cognitive of the Aristotelian ideas concerning abduction in (Magnani, 2015b, 2016, 2017) and in the recent (Magnani, 2021).

[2] Further details concerning the EC-model of abduction can be found in (Magnani, 2015b, 2016).

knowledge enhancing role of abduction, that is when abduction is not considered an inference to the best explanation (IBE) in the classical sense of the expression, that is an inference necessarily characterized by an empirical evaluation phase, or inductive phase. In chapter three of (Magnani, 2017) I have provided the example of conventions: abducing *conventions* favors and increases knowledge even if these hypotheses remain evidentially inert—at least in the sense that it is not possible to empirically falsify them. Consequently, abduced conventions are evidentially inert but knowledge enhancing at the rational level of science.

A rich analysis of the various criteria (or—more simply—of the various intermediate criteria for provisionally "judging" abductive results, such as plausibility, relevance, consistency, simplicity, minimality, likeliness, loveliness, probability, uberty, fruitfulness, etc.) that characterize a "best explanation", as they are illustrated by the rich literature on the issue produced in last decades, is offered by (Cabrera, 2017, 2020). Cabrera proposes a less misleading name for IBE: "inference to the hypothesis with the optimal combination of explanatory virtues". This qualification emphasizes the contextual character and the fact that a privileged model of explanation is irrelevant and unnecessary. It is also usefully observed: "Of course, explanationists must tackle other obstacles, such as offering more detailed accounts of the virtues, of why those virtues are epistemically relevant, and of how the virtues are to be agglomerated and compared. Still though, appealing to the explanatory virtues [...], provides, in my view, an elegant solution to what would otherwise be a devastating objection to IBE" (Cabrera, 2020, p. 21).

Assessing that there is a common ground in all these works of what could be broadly defined as "creativity" does not imply that all of these forms of selective or creative abduction[3] with their related cognitive strategies are the same, contrarily it should spark the need for firm and sensible categorization: otherwise it would be like saying that to construct a doll, a machine-gun and a nuclear reactor are all the same thing because we use our hands in order to do so!

## 1.3 Abduction and the Optimization of Eco-Cognitive Situatedness

When I say that abduction can be knowledge-enhancing I am referring to various types of newly produced knowledge of various novelty levels, in absence of an empirical evaluation phase, or inductive phase, as Peirce called it. Some cases of new knowledge produced in science (for example, conventions and intermediate

[3] For example, selective abduction is active in diagnostic reasoning, where it is merely seen as an activity of "selecting" from an encyclopedia of pre-stored hypotheses; creative abduction instead refers to the building of new hypotheses. I have proposed the dichotomic distinction between selective and creative abduction in (Magnani, 2001). A recent and clear analysis of this dichotomy and of other classifications emphasizing different aspects of abduction I have described is given in (Park, 2015).

models used in research settings), are cases of knowledge enhancing abduction. However, also knowledge produced in an artificial game thanks to a smart application of strategies or the invention of new strategies and/or heuristics has to be seen as the fruit of knowledge-enhancing abduction. This means that abduction is not necessarily ignorance-preserving (reached hypotheses would always be "presumptive" and to be accepted they always need empirical confirmation), as contended by Gabbay and Woods (2005), with their Gabbay-Woods schema of abduction (GW-schema). Abduction can creatively build new knowledge by itself (that is as an inference not necessarily characterized by an empirical evaluation phase, or inductive phase), as various examples coming from the area of history of science and other fields of human cognition clearly show.[4]

I contend that to reach selective or creative good abductive results efficient strategies have to be exploited, but it is also necessary to count on an environment characterized by various degrees of what I have called *eco-cognitive situatedness*, in which that eco-cognitive openness already envisaged by Aristotle thanks to the emphasis on "leading away"[5] is fundamental (Magnani, 2016). To favor good creative and selective abduction reasoning strategies must not be "locked" in an external restricted eco-cognitive environment such as in a scenario characterized by fixed definitory rules and finite material aspects (an artificial game, Go or Chess, for example), which would function as cognitive mediators able to constrain agents' reasoning.[6] In brief, the optimization of eco-cognitive situatedness concerns the substantial problem of *discoverability* and *diagnosticability*, almost totally disregarded in the literature on abduction (and just sketched by Peirce himself). I have to anticipate that the solution of this problem has to take into account the fact that we deal with something fundamentally characterized by contextual configurations.

Research on abduction has frequently emphasized the fruitful role of cognitive openness. (Hendricks & Faye, 1999) consider trans-paradigmatic abduction a form of discovery in which a guessed hypothesis transcends the prompt empirical agreement between two paradigms. The paradigms are presumed to belong to the same field (for example physics) where one of the fields is well established and the other is emerging (for example classical and quantum physics):

> New theoretical concepts can be introduced transcending the current body of background knowledge while yet others remain within the given understanding of things. A case in point would be the formulation of the hypothesis of electron spin. Bohr considered the spin conjecture as a welcome supplement to the current magnetic core theory. Pauli remained rather skeptical pertaining to the spin hypothesis due to the fact that it actually required the theory of quantum mechanics for its proper justification, which was not part of the

[4] I better supported my claim about the knowledge enhancing character of abduction in the recent (Magnani, 2015b, 2016). Woods has recently enriched, modified, and moderated his views of ignorance-preservation, see (Woods, 2017).

[5] Cf. the previous section of this chapter.

[6] More details concerning the role of locked and unlocked strategies are illustrated in the recent (Magnani, 2019) and in chapter three of this book.

> background knowledge at the time of the conjecture. In such cases, two paradigms are competing and the abduction is then dependent upon whether the conjecture is made within the paradigm or outside it. Hence we distinguish between paradigmatic and trans-paradigmatic abduction (cit., p. 287).

Furthermore, people draw on different domains of knowledge to arrive to an abductive conclusion thanks to what (Gibson & Bruza, 2021) call "transepistemic abduction" (TeA), which illustrates how two agents, in order to successfully explain a phenomenon, reason across two very distant cognitive fields (for example computational and psychosocial domains) despite each agent being ignorant of the other domain knowledge. The authors themselves acknowledge that TeA represents a case that is partially concerned with my eco-cognitive perspective: "TeA may not necessarily accommodate wider understandings of abduction like the eco-cognitive model proposed by Magnani. For example, TeA may not necessarily encompass perceptions, aesthetic decisions or moral judgements in the way that a eco-cognitive view of abduction might" (p. 475).

Another interesting procedure that can refer to higher abductive processes in need of cognitive openness is the chunk-and-permeate method (Brown & Priest, 2004), in which consideration is given to conditions under which mutually incompatible well-grounded theories can interact to bring forth solutions to problems which neither theory can solve on its own. This method introduces a paraconsistent reasoning strategy, in which information is broken up into chunks, and a limited amount of information is allowed to flow between chunks, and it is applied to model the reasoning employed in the original infinitesimal calculus.

## 1.4 Irrelevance and Implausibility Defended

From the perspective of the EC-Model the available logical views concerning abduction have to be reconsidered. By now this fact is acknowledged by some logicians. Estrada-González (2013, p. 182), for example, by referring to the properties called *consistency*, *minimality*, etc. stated in the AKM-schema[7] usefully notes:

> We think that overemphasizing the characteristics of abduction as it occurs in scientific practices and daily life scenarios has led to overlook some features that abduction in those circumstances shares with other phenomena in which some given outputs fail to stand in a certain relation with some given inputs and thus a modification on those inputs is in order.

Indeed, abduction in an eco-cognitive perspective is not circumscribed by indicating particular requirements of each particular field of application, such as scientific discovery, diagnosis, machine discovery in artificial intelligence (AI), and

[7] Gabbay and Woods (2005) call $AKM$-schema: for $A$ they refer to Aliseda (1997, 2006), for $K$ to Kowalski (1979), Kuipers (1999), and Kakas et al. (1993), for $M$ to Magnani (2001) and Meheus et al. (2002), which is contrasted to their own ($GW$-schema), which I have explained in chapter two of my book (Magnani, 2009) and below in this chapter.

computational creativity. The available logical models of abduction (orchestrated around the properties of classical deduction), the so-called nonmonotonic, adaptive, etc. logics, which restrict their structure to some of the above requirements, are extremely important and provide images of abduction that, even if logically powerful, are partial, limited, and not appropriate to represent general perspectives.

Of course, we acknowledge that the rich research on abduction already available in the fields of logic, cognitive science, philosophy, and artificial intelligence (AI), has already rehabilitated the "cognitive" importance of the fallacy "affirming the consequent" (abductive reasoning corresponds to this fallacy in the light of classical logic), traditionally taken as the mistake of having a conditional and its consequent, and from this deriving the antecedent. When reframed in the spirit of the naturalization of logic[8] this fallacy becomes a form of abduction endowed with a positive cognitive value, in most of the real-life reasoning contexts in which it occurs, included diagnosis and creative processes. It can exhibit what some writers have called "material validity",[9] in this case to refer to the fact that an invalid form provides a cognitive good semantic outcome. We need a new logical framework for abduction: an important step of the naturalization of this fallacy is certainly the already widespread acknowledgment of the *agent-based nature* of abduction but also—as I will stress below, taking advantage of my EC-Model—of its *eco-cognitive situatedness*.

The already available taxonomy of the various kinds of abduction, which myself and many authors have proposed and analyzed, is extremely useful because it shows that we are dealing with cognitive activities extremely variegated but also tremendously fruitful, resorting to the common aim of guessing hypotheses that can provide various kinds of reliable knowledge, common, scientific, moral, aesthetical, etc. These abductively established kinds of cognition do not necessarily aim at providing truths in the sense of the word established by the tradition of the epistemology of natural sciences: for example, abduction certainly works when performed to establish a new "truth" in physics, but also when exploited to make a "good" hypothesis (for example about a person) in gossip. In this last case the reaching of a truth based on evidence, that is a truth that would be such because, for example, in tune with the spirit of scientific rationality—is usually a secondary concern. For example, in the case of cognitive moral interplays of gossip, it is the produced hypothesis (about a person or an event) that establishes—constitutes, we could say—a special kind of "truth" by itself, that truth that is

[8] In (Magnani, 2015a) I have illustrated and commented on Woods' program of naturalization of logic advanced in his recent book *Errors of Reasoning: Naturalizing the Logic of Inference* (Woods, 2013). This program aims at bringing logic into a creative rapprochement with cognitive science. This can be achieved by trying to do for logic what over forty years ago Quine and others attempted for epistemology: similarly, it is necessary to propose a "naturalization" of the logic which leads logic closer to actual human inference through a process of de-idealization.

[9] The standard concept of material validity is illustrated in (Brandom, 2000), as a case of semantically valid inference, which instantiates an invalid syntactic form. I adopt the same term slightly modifying its meaning.

such because locally/situationally accepted and efficiently adopted by the involved human group: this is just what it takes, to promote the narratives or the actions—sometimes violent—which the relationships between the individuals of the group aim at performing.[10]

First of all, we have to say that an abductive inference is not different from other inferences: it presents outputs derived from inputs: the AKM and GW schemas I have quoted above treat abduction in this way. A special perspective can be assumed, which shapes the inferential status of abduction in a smart way. I can anticipate that, in this perspective, abduction appears orthogonal to deduction, but not incompatible with it, such as it is occurring when abduction is seen as a fallacy, in the light of the classical Peircean syllogistic framework: the important difference is indeed that in the perspective I will soon illustrate (see the following section) no direct reference to the fallacious character of abduction is postulated. Hence, to the aim of "naturalizing" abduction in logic, and to follow the main tenets of my EC-Model, it is provisionally appropriate to adopt a broad logical view of the so-called *inferential problems*, in which the semantic problem of the *fallacious* character of abduction and the problem of its *presumptive* character (stressed by the GW schema) are not primarily taken into account.[11]

Let us adopt, following Estrada-González (2013) the more general concepts of input and output instead of premisses and conclusions: as I have already explained at the beginning of this chapter, this view consists in seeing inference not only in terms of the reaching of an output or the modification of it, like in the standard view of deductive proofs,[12] but also as the *process of modifying part of the inputs, or both inputs and outputs*, in order to obtain the desired relation between inputs and outputs. In this framework, abductive cognition is easily depicted in terms of the situation in which some given outputs fail to stand in a certain relation with some given inputs and thus a modification on those inputs is needed. It is a received and common way of modeling abduction in logic (cf. for example Aliseda, 2006).

Still following Estrada-González, let us consider a simple example such as the consequence relation of a logic $L$, $\Vdash_L$, and let $\alpha$, $\beta$ be formulas of the language of $L$, © is a certain binary connective. An abductive problem can be illustrated by the following incomplete argument, in the sense that the conclusion does not $L$-follow

[10] More details about hypothetical cognition in gossip are illustrated in Magnani (2011) and Bertolotti & Magnani (2014).

[11] Anyway, we have to remember that the presumptive character of abduction manifestly stresses that necessity—taken in the received Aristotelian sense—is just only one criterion which can characterize the solution of an inferential problem, if we consider it from a wide point of view. Also plausibility, probability, possibility, sufficiency, defeasibility are completely legitimate (Woods, 2013, chapter eight). A general inferential problem is not inevitably characterized by the "necessity" of classical deduction, in the Aristotelian sense, other aspects have to be taken into account and considered legitimate.

[12] From this point of view, deductive and inductive inferences are situated at the same level because they concern the finding of suitable modifications in the outputs (Estrada-González, 2013, pp. 189–191).

from the premisses

$$\alpha \copyright \beta \nVdash_L \beta \qquad (1.1)$$

The making of an abduction I have just described represents the logical counterpart of the cognitive (and epistemological) *fill-up problem*[13] I have already quoted above, that is the finding of a suitable improvement of the context (the premisses), here the variable ?, such that from the original context and its enrichment the conclusion $L$-follows

$$\alpha \copyright \beta, ? \Vdash_L \beta \qquad (1.2)$$

Obviously, there are many ways to complete the argument above. A list of various presumptive very elementary solutions can be easily provided (for example $\alpha$, or $\beta$), showing that to the aim of choosing among them we need some preferential measure: this resorts to the *cutdown* problem.[14]

[13] "There is the problem of finding criteria for hypothesis *selection*. But there is the prior problem of specifying the conditions for *thinking up* possible candidates for selection. The first is a 'cutdown' problem. The second is a 'fill-up problem'; and with the latter comes the received view that it is not a problem for logic" (Woods, 2011, p. 243 emphasis added). In chapter six of my recent book on abduction (Magnani, 2017) I have contended that in a wide eco-cognitive perspective the cutdown and fill-up problems in abductive cognition appear to be spectacularly *contextual*.

[14] In the scheme (1.2), according to Estrada, we might need to add rules and the conclusions may even be rules. What is needed is an input. However, the scheme represents a kind of oversimplification: abducing new rules cannot simply consist in adding an input instead of ?. It rather requires some kind of dynamic of the relation expressed by $\Vdash_L$. As I will better explain below, in the case in which new rules are abduced, what is changed is the relation between inputs and outputs, not only the inputs (even if the new rule is some kind of input as well). This is for example highlighted by considering a dynamic reading of $\Vdash_L$, and that should be accounted for by the scheme (1.3) at p. 16, in which $X$ in $\Vdash_L^X$ expresses different potential relations between input and output and $\Vdash_L^{\surd}$ expresses the fact that—dynamically—the expected relation is reached. I repeat, in the simple scheme (1.2), what is needed is clearly an input. However, abducing new rules cannot simply consist in adding an input instead of ?. This is a point also addressed by (Barés-Gómez & Fontaine, 2021) and (Fontaine & Barés-Gómez, 2019) in their adaptive dialogical approach to abduction: in dialogues, there are two kinds of rules. The particular rules provide the local semantics of the connectives. The structural rules define the context of argumentation (the underlying logic), thereby the global semantics of the connectives. A change of structural rules involves a change of logic. By articulating different sets of structural rules in a specific kind of game, a dynamic consequence relation may be obtained. This is precisely how Fontaine and Barés conceive abduction of rules in dialogues: while the argumentative partners may be interacting in the context of a dialogue for paraconsistent logic (with constraints on the use of negation), the hypothesis that negation behaves consistently at some point in the dialogue may be conjectured (together with a certain commitment to defend further conditions). This might appear less general, but generality is here to be understood in terms of the structure of the dialogues which can be similarly adapted to other dynamic processes. Hence, this perspective approaches abduction without disregarding its pragmatic/dialectical dimension, thanks to the adoption of a dialogical logic. This current logical illustration of the dialectic involved in abduction is able to model argumentative interactions leading to conjectures, through a kind of "dialectification".

It is clear that in principle there might be infinitely many ways to complete the above argument. Estrada-González (2013, p. 183) notes that abduction "[...] is usually thought as the construction of a (scientific) hypothesis for adding it to a theory and other previous knowledge (the premises) in order to explain a problematic phenomenon (the conclusion)". If we adopt this received schema of abduction in our simple example various solutions should be discarded because they are not good options: for example, according to some chosen criteria, we can say that solutions do not have to be trivial, redundant: for example, if one has a connective $K$ such that $\beta$ is $L$-implied by both $(\alpha©\beta)K\beta$ and $\alpha©\beta$, the solution can obviously be to add the formula $(\alpha©\beta)K\beta$. Also, they do not have to lack explanatory power (such as for example the solution $\beta$). Furthermore, a solution that is not simple, and so excessively explanatory, certainly less simple than another available, has to be discarded, such as in the following cases, in which complicated solutions are added: $\alpha©\beta, \vartheta, \beta, \alpha, (\alpha©\beta)K\beta \Vdash_L \beta$ or $\alpha©\beta, (\alpha©\beta)K\beta, \Vdash_L \beta$. Indeed, there is another solution available that is the preferred one—the solution $\alpha$.

In sum, by adopting the received logical view of abduction (AKM-schema), we restrict our general idea of abduction being conditioned by special cases (such as the dominant ones of medical diagnosis): unfortunately, by restricting our perspective in this way, some good solutions that would be remarkable and productive in other cases (for example different with respect to scientific discovery and medical diagnosis) can be ruled out. To avoid this result the GW-schema and the EC-Model of abduction can be of some help. The GW-schema does not refer to consistency and minimality as necessary requirements.

However, it has to be said that many standard perspectives on abduction still demand two properties, which are presented as possessed by "every" kind of solution for an abductive problem:

1. *Relevance*: the solution, the guessed hypothesis $H$, should be relevant to the problem: for example, if an agent's knowledge does not suffice to know why the bartender in Kuala Lumpur has been killed, releasing the true Newtonian hypothesis that the planets move according to the law of gravitation has nothing to do with the given problem: it is not relevant.
2. *Plausibility*: The abduced hypothesis $H$ should be characterized by some designated degree of plausibility. If an agent's knowledge does not suffice to know who killed the bartender in Kuala Lumpur, releasing the hypothesis that the killer is the President of United States has to do with the problem (because, after all, the President is a human being and we know human beings are potential killers) but is sufficiently implausible as to count as a solution.

Since for many abductive problems there are—usually—many guessed hypotheses, the abducer needs to reduce this space to one. This means that the abducer has to produce the best choice among the members of the available group: It is extremely difficult to see how this is done, both formally and empirically. There is the problem of finding criteria for hypothesis *selection*. But there is the prior problem of specifying the conditions for *thinking up* possible candidates for selection, I have already explained above. In particular, formal models aim at producing fragments of classical logic in which instances of abductive or backward reasoning are allowed. It is also interesting to note that recent studies, even if not directly related to the naturalization of the logic of abduction, are strongly concerned with the role of context in reasoning: Zardini emphasizes, in a recent rich formal treatment, the role of simple temporal "intercontextual" logics, which aim at adequately modeling the validity of certain arguments in which the context changes (Zardini, 2014).

Indeed in the eco-cognitive perspective the relevance of a guessed hypothesis would seem a trivial requirement because it is "hard to see how it might fail to be relevant", as Estrada-González says (2013, p. 185). He further adds: "Someone might press the point that what is required is the relevance not of solutions, but of candidates to be solutions. However, I think it might go against all those pleas connecting abduction with creativity, hypothesis generation, guessing, etc." I basically agree with him. This is exactly the point to be stressed and further explained. First of all, we have to note that relevance is *context-* and *time-dependent*. When Feyerabend (1975) emphasizes the role of what he calls "counterinduction", he is just presenting to us the complete unreasonable and unwarranted character of scientific discovery: the guessed hypothesis could be devoid of relevance to the problem in the framework of the upholders of the rival theory but also, even if not necessarily, in the perspective of the agent herself that—paradoxically—guessed the new "strange" hypothesis. Of course, the relevance requirement is related to the current state of knowledge of both agonists. However, the new hypothesis can result "relevant" later on, for example when recognized as a new discovery. To summarize, candidates to be solutions which seem weird—irrelevant—soon can become relevant if they are recognized as solutions.

Something similar can be said in the case of *plausibility*. First of all, in general, we cannot be sure that our guessed hypotheses are plausible (even if we know that looking for plausibility is a human good and wise heuristic), indeed an implausible hypothesis can, later on, result plausible. Moreover, when a hypothesis solves the problem at hand, this is enough to count as a solution of the abductive problem (even if, not necessarily a *good* solution or the *best* solution). If we want to preserve the property of plausibility, at most, we can say that in some cases it is just *potential*, given the time-dependency I have illustrated.

To make an example, the strange Cartesian hypothesis of a plenum vortices made of particles, destroyed by the Newtonian concept of action at distance, later on, appeared more rational and fully compatible with the Einsteinian framework:

> Thus Descartes was not so far from the truth when he believed he must exclude the existence of empty space. The notion indeed appears absurd, as long as physical reality is seen exclusively in ponderable bodies. It requires the idea of the field as the representative of reality, in combination with the general principle of relativity, to show the true kernel of Descartes' idea; there exists no space "empty of field" (Einstein, 2014, pp. 375–376).

In sum, irrelevance and implausibility not always are offensive to reason:[15] to delineate the fill-up problem neither relevance, nor plausibility are necessary, they are just two "typical" smart and fruitful principles human beings subjectively adopt to look for hypotheses. Unfortunately, they are no longer "typical", for example, in the case of high-level kinds of cognitive creativity. Also the GW-schema acknowledges the fact that relevance or plausibility cannot be taken to be general conditions of hypothesis selection.

It is now useful to provide a short introduction to the concept of eco-cognitive openness from a logical point of view. The new perspective inaugurated by the so-called *naturalization of logic*[16] contends that the normative authority claimed by formal models of ideal reasoners to regulate human practice on the ground is, to date, unfounded. It is necessary to propose a "naturalization" of the logic of human inference. Woods holds a naturalized logic to an adequacy condition of "empirical sensitivity" (Woods, 2013). A naturalized logic is open to study many ways of reasoning that are typical of actual human knowers, such as fallacies, which, even if not truth preserving inferences nonetheless can provide truths and productive results. Of course one of the best examples is the logic of abduction, where the naturalization of the well-known fallacy "affirming the consequent" is at play. (Gabbay & Woods, 2005, p. 81) clearly maintain that Peirce's abduction, depicted as both a) a surrender to an idea, and b) a method for testing its consequences perfectly resembles central aspects of practical reasoning but also of creative scientific reasoning.

It is useful to refer to my recent research on abduction (Magnani, 2016), which stresses the importance in good abductive cognition of various kinds of the already quoted *optimization of situatedness*: abductive cognition in a situation of "strong" eco-cognitive openness is for example very important in scientific reasoning because it refers to that activity of creative hypothesis generation which characterizes one of the more valued aspects of rational knowledge. I think that in scientific creative reasoning discoverability is only granted by a *maximization/optimization of eco-cognitive openness*. The study above teaches us that situatedness is related to the so-called eco-cognitive aspects, referred to various contexts in which knowledge is "traveling":[17] to favor the solution of an inferential problem—especially in science but also in the case of other abductive problems, such as diagnosis—the richness of the flux of information has to be maximized.

In my opinion, it is exactly the presence of a "maximization/optimization of eco-cognitive openness" that can favor the discoverability of abductive hypotheses characterized by what Peirce called "uberty", as a potentiality to arrive at undis-

---

[15] Scientists would agree that the really surprising and fruitful thought arising from abduction has to challenge prevailing conceptions by suggesting ideas that are *prima facie* neither relevant nor plausible, nor that even appear to contradict pre-established notions (see e.g. the book by Livio, 2013).

[16] I have illustrated this new project in (Magnani, 2015a). See also above the last part of Sect. 1.2.

[17] Cf. chapter three, this book.

covered truths: "I think logicians should have two principal aims: 1st, to bring out the amount and kind of security (approach to certainty) of each kind of reasoning, and 2nd, to bring out the possible and desirable uberty, or value in productiveness, of each kind" (Peirce, 1866–1913, 8.384). As I have illustrated above hypotheses of this kind—uberous hypotheses are a typical possible fruit of abduction—are often irrelevant and implausible in the perspective of the contemporary context but have a high potential to favor undiscovered truths. This means that plausibility and relevance are often deceptive guidance that can generate a bad result: the overlooking of potentially fecund hypotheses. The price of uberty is lost security.[18]

## 1.5 Identifying the Optimization of Eco-Cognitive Situatedness in a Logical Perspective

A good abductive logical system is classically characterized by the following general distinct levels, which resort to the fill-up and cutdown aspects I have already explained above:

- a *base logic* $L_1$ with proof procedures $\Pi$;
- an *abductive algorithm* which deploys $\Pi$ to look for missing premisses and other formulas to be abduced;
- a *logic* $L_2$ for deciding which abduced formulas to choose, which criteria of selection apply, etc. This logic is related to the specification of suitable constraints regarding plausibility, relevance (topical, full-use, redundancy-oriented, probabilistic), etc., and economy, making the ideal agent able to discount and select information that does not resolve the task at hand (Gabbay & Woods, 2005).

As I have illustrated above it is thanks to the so-called GW-schema that Gabbay and Woods criticize what they call the classic AKM-schema of abduction. A primary gift provided by this new GW-schema was the opening of the discussion about *ignorance preservation* but also about *non-explanatory* and *instrumental* abduction, considered as not intrinsically consequentialist.[19]

In the previous parts of this chapter, I have underlined that relevance and plausibility are *context-* and *time-dependent*, and so they are not necessary aspects of all potential successful abductions: this means that also in this perspective eco-cognitive environmental situatedness of abductive cognition matters more than expected. From the perspective of logic, the first consequence is that the fill-up problem should consist of a logic *eco-cognitively disciplined* and *multimodally* built. I have indicated above—also taking advantage of the analysis of the Aristotelian

[18] On Peirce and uberty cf. (McJohn, 1993; Campos, 2011; Mcauliffe, 2015; Pietarinen, 2020).

[19] I have illustrated non-explanatory and instrumental abduction in (Magnani, 2009, chapter two), also providing some case studies.

concept of "leading away" (ἀπαγωγή)—that looking for a naturalization of the logic of abduction, it is fruitful to follow the main tenets of my EC-Model. Consequently it is appropriate to adopt, when dealing with the so-called "inferential problem" of abduction, the more general concepts of input and output instead of those of premisses and conclusions, in the perspective I have already illustrated above. In such a way there is one more advantage: it is more natural to accept the *multimodal* character of the inferences involved. In this perspective we expect to have a logic in which the artificial language presents an *extendable* expressive capacity and that can also be itself composed of icons and not only of symbols, like it is occurring in the case of heterogeneous logics. Some logicians seem to acknowledge this need when dealing with a broad view of the inference problems: "An inference can be seen as an argument completion process, where premises are reduced either to a single formula or to multiple formulas, depending on whether single—or multiple—conclusion logics are used. But although the conclusion might consist of a single formula, we might need to add rules, theories, etc. to complete an argument, and not just adding formulas in the context. Nonetheless, the conclusions could be also rules (if for example, in a logic we want a rule to be a derived one, we may need to make some input to the original logic), theories, valid arguments, diagrams, etc." [...] (Estrada-González, 2013, p. 186).

To the aim of explaining and describing how discoverability and diagnosticability are so important in abductive cognition the reader has to be patient and to examine the following characterization of abduction, which will highlight and illustrate the problem of optimization of eco-cognitive situatedness, taking advantage of simple logical considerations. Let $\Theta = \{\Gamma_1, \ldots, \Gamma_m\}$ be a background theory, $P = \{\Delta_1, \ldots, \Delta_n\}$ a set of true sentences corresponding—for example—to phenomena to be explained and $\Vdash$ a consequence relation, usually—but not necessarily—the classical one. According to the received AKM schema I have quoted above, abduction refers to the inference of $P$ from $\Theta$ and $H = \{A_1, \ldots, A_k\}$, a collection of hypotheses, given some further constraints, which can basically relate to consistency, minimality, and preference, as we have repeatedly stated. In this perspective an abductive problem concerns the finding of a suitable improvement of $A_1, \ldots, A_k$ such that $\Gamma_1, \ldots \Gamma_m, A_1, \ldots, A_k \Vdash_L \Delta_1, \ldots, \Delta_n$ is *L-valid*. It is obvious that an improvement of the input $\Gamma_1, \ldots \Gamma_m$ can be reached by additions of a "new" background theory but also by modification and enrichment of the input already available in the given inferential problem: this process refers to the importance of the "optimal position" of both input and output, as I will better describe in the following section.

In the perspective above in terms of inferential problems an inference is not only seen in terms of the process that leads to the generation of and output or to the proof of it, like in the traditional and standard view of deductive proofs, when we have to obtain the output from the input. In this broader logical view an inferential problem is also the process of *increasing* or *modifying* part of the input, or both *input and output*, in order to obtain a desired relation between input and output. Of course, the reached extension or modification is the solution of the inferential problem. It is clear that in the case of abductive cognition input and output fail to stand each other in an expected relation and the solution requires enrichment of part of input, not of output: these enrichment represents the solution of the abductive problem.

I have already said that this process of modification of input is basically multimodal and various context-depending conditions have to be fulfilled in order to reach the (best) solution. From this perspective the general form of an inferential abductive problem can be symbolically rendered as follows (Estrada-González, 2013, p. 189):

$$\Lambda_1, \ldots, \Lambda_i, ?_I \Vdash_L^X \Upsilon_1, \ldots, .\Upsilon_j \quad (1.3)$$

in which $\Vdash_L^X$ indicates that input and output do not stand each other in an expected relation and that the generation of the input $?_I$ can provide the solution. $\Vdash_L^{\surd}$ will denote that the expected relation is obtained.

In general, in this characterization, the direction is not from evidence/premisses to abductive result but the forward fashion is adopted, where the inferential parameter $\Vdash$ sets some appropriate logical relationship between an input which consists in both the abductive guess to be found and a background theory, and an output—premisses, for example, an evidence, a novel phenomenon to be abductively "explained" through facts, rules, or even new theories. The inferential parameter does not have to be considered neither the semantic entailment above nor the classical derivability. Aliseda observes that it ranges—at least in the case of the main received non-standard logical accounts—over diverse values such as for example probable inference, logic programming, or dynamic inference: "abduction is not one specific non-standard logical inference mechanism, but rather a way of using any one of these" (Aliseda, 2006, p. 47); also, she stresses that it is logically useful to depict "explanatory" consistent abduction by extracting those sets of "top-down" (van Benthem, 2007, p. 272) basic properties that are called *structural rules* of inference, such as conditional reflexivity, simultaneous cut, conclusion consistency, modified monotonicity, modified cut, rejection of permutation in dynamic inference, etc., which "fit a logical format" (Aliseda, 2006, p. 150).

The structural rules reverberate—Aliseda (cit., p. 95) says they "state how abductive explanatory logic behaves"—the fact that abduction can be expressed by deviant (but still logical) systems, endowed with extra-systematic specific notions of validity, even if, unfortunately, they do not provide methods for *generating* abductive explanations, such as it is occurring computing abduction thanks to logic programming and semantic tableaux.

Finally, it has to be remembered that, if we consider the generation of the input $?_I$ that provides the solution as *subjunctively* attained, such as in the case of the ignorance-preserving abductions illustrated by the GW-schema, then these computing devices do not appear to be "genuine" abductions. In this perspective Woods observes that in the case of semantic tableaux abduction resembles enthymeme resolution and so does not reflect its presumptive character: the task resorts to the rehabilitation of an ailing deduction. In semantic tableaux abduction, "the task is to find a $\phi$ that closes a *model*-connection between a theory and some (usually) empirical data, which also is the repair of a decrepit deduction" (Woods, 2007, p. 310). In the light of the *presumptive* character of the GW schema, and only in this light, there is nothing that is abductive about such closures. Indeed $\models$-closures are possible without there being any ignorance-problem to which the closure is a response.

### *1.5.1 The Importance of "Optimally Positioning" Input and Output*

I contend that to get good abductions, such as for example the creative ones that are typical of scientific innovation, the input (background theory) and output (evidence, premisses) of the formula $\Lambda_1, \ldots, \Lambda_i, ?_I \Vdash_L^X \Upsilon_1, \ldots, .\Upsilon_j$ (in which $\Vdash_L^X$ indicates that input and output do not stand each other in an expected relation and that the generation of the input $?_I$ can provide the solution) have to be thought as *optimally positioned*. Not only, this optimality, in various types of abduction even if not in all, is made possible—at least in scientific discovery—by a *maximization of changeability* of both input and output; again, not only input has to be enriched with the possible solution but, to do that, other aspects of input have usually to be changed and/or modified.[20]

Indeed, in our eco-cognitive perspective, an "inferential problem", given some starting evidence/premisses, can be enriched by the appearance of a new output (that is more evidence/premisses) to be accounted for and the inferential process has to restart. This is exactly the case of abduction and the cycle of reasoning reflects the well-known nonmonotonic character of abductive reasoning. Abductive consequence is ruptured by new and newly disclosed information, and so defeasible. In this perspective abductive inference is *not only* the result of the generation of the input $?_I$ (that is the reaching of the abductive hypothesis), but, in general, actually involves the intertwined modification of both input (also the background theory to be adopted) and output (the evidence/premisses). Consequently, abductive inferential processes are highly *information-sensitive*, that is the flux of information which interferes with them is continuous and systematically human(or machine)-promoted and enhanced when needed. This is not true of traditional inferential settings, for example proofs in classical logic, in which the modifications of the input are *minimized*, proofs are usually taken with "given" input and the burden of proofs is dominant and charged on rules of inferences, and on the smart choice of them together with the choice of their appropriate sequentiality. This changeability first of all refers to a wide psychological/epistemological openness in which (at least in science) knowledge transfer has to be maximized.

In sum, considering an abductive "inferential problem" as symbolized in the above formula, a suitably "anthropomorphized" logic of abduction has to take into

[20] More details are illustrated in (Magnani, 2016, section three).

account a continuous flux of information from the eco-cognitive environment and so the constant modification of both input and output on the basis of both

1. the *new information available*,
2. the *new information inferentially generated*, for example new inferentially generated input aiming at solving the inferential problem.

To conclude, optimization of situatedness is the main general property of logical abductive inference, which—from a general perspective—defeats the other properties such as minimality, consistency, relevance, plausibility, etc. These are special subcases of optimization, related to the quality of the guessed hypotheses, and are intertwined with the kind of a given situatedness, at the level of the appropriate abductive inference to the new input indicated with a question mark of the above formula.

Another count is indeed the problem of choosing, selecting, and finding the appropriate input (background theory) and output (evidence/premisses), that is an optimal positioning of the eco-cognitive situatedness, to the aim of "favoring" the required kind of abductive process.

We have seen that an abductive solution is still related to that Aristotelian "leading away" (ἀπαγωγή)—quoted above in Sect. 1.2—that is on the starting of the application of a *supplementary logic* implementing an appropriate formal inference engine which can provide a solution. This supplementary logic implements a new inference engine from the output of the formula (1.3). If we want to naturalize the logic of the abductive processes and its special *consequence relation*, which has to be strongly "eco-cognitive-sensitive", $\Vdash_{EC}$, we must at this point note that a naturalized abductive logic would have to first of all refer to the

- *optimization of situatedness*. Situatedness is related to eco-cognitive aspects: to favor the solution of the abductive problem input and output of the formula above have to be thought as *optimally positioned*.

Other accessorial aspects which favor the success of abductive cognition have to be immediately illustrated:

1. in scientific discovery (but also in other fields of abductive cognition, such as for example medical diagnosis) the above-illustrated optimality has to be further characterized as a *maximization of changeability* of both input and output. Not only inputs have to be enriched with the possible solution to favor the potential increase of knowledge, even if not satisfactory, so favoring the restarting of a new inferential process, but, to do that, other inputs (background theory) have usually to be built, changed, and/or modified;
2. consequently, abductive inferential processes are highly *information-sensitive*, that is the flux of information which interferes with them is continuous and systematically human(or machine)-promoted and enhanced when needed. This is not true of traditional inferential settings, for example, proofs in classical logic, in which the modifications of the input are *minimized*, proofs are usually taken with "given" input, and the burden of proofs is dominant and charged on rules

of inferences, and on the smart choice of them together with the choice of their appropriate sequentiality;
3. indeed, in our eco-cognitive perspective, an "inferential problem" can be enriched by the appearance of a new output to be accounted for and the inferential process has to restart. This is exactly the case of abduction and the cycle of reasoning reflects the well-known nonmonotonic character of abductive reasoning. Abductive consequence is ruptured by new and newly disclosed information, and so defeasible. In this perspective abductive inference is *not only* the result of the modification of the input, but, in general, actually involves the intertwined modification of both input and output;
4. contrarily to the case of classical demonstrative systems, which are characterized by what I call *maximization of memorylessness*, naturalizing a logic of abduction is related to the need of keeping record of the past life of abductive inferential praxes;
5. traditional logical systems are *abstract* in the sense that they are based on a *maximal independence* regarding sensory modality, and so they strongly stabilize experience and common categorization, on the contrary, this requirement is no longer crucial, but "abstractness" is circumstance-based and so eco-cognitively conditioned. Moreover, I repeat, multimodality of formalization is really important: a naturalized logic of abduction is open to a modification of both language and rules;
6. as Peirce says, abduction is "akin to the truth":[21] research on abduction in the last decades certainly has described it as an exceptional example of a significant and fruitful truth-generating non-deductive reasoning.

Further, let us remember that in an abductive "inferential problem" as symbolized in (3) (cf. above, Sect. 1.5), it is extremely important to grant *multimodality*. The logical inferential process that is involved in the whole modification of input and output has to be strongly intended as *multimodal* not only from the point of view of the *cognitive devices* "represented" (not merely propositions, but also diagrams, for example) but also from the point of view of the *applied rules*: model-based, but also computational aspects, have to be taken into account.

## Appendix: Abduction as the Eco-Cognitive Production of a Deduction

In Sect. 1.4 above we have remembered that the making of an abduction represents the logical counterpart of the cognitive (and epistemological) *fill-up problem*, that is the problem of finding a suitable improvement of the context (the premisses)—

[21] "It is a primary *hypothesis* underlying all abduction that the human mind is akin to the truth in the sense that in a finite number of guesses it will light upon the correct hypothesis" (Peirce, 1866–1913, 7.220).

taking advantage of what I have called eco-cognitive openness—and finally the variable ? below, such that from the original context and its enrichment the conclusion $L$-follows

$$\alpha \copyright \beta, ? \Vdash_L \beta \tag{1.4}$$

This abductive need of improving the context (the premisses) echoes what is described in chapters A27–A30 of the first book of *Prior Analytics*,[22] that is the group of procedures—we could now call heuristic—for finding premises to decently solve certain problems, Aristotle says " τὸ ποιεῖν συλλογισμὸν " (Aristotle, 1989, A28, 43a, 19–24, p. 42), that is to the aim of *producing* a syllogism in a satisfactory way. Indeed, at the beginning of Chapter A27 Aristotle contends that "Now it is time to explain how we may ourselves always be supplied with deductions about what is set up, and the route by which we may obtain the principles concerning any particular subject. For surely one ought not only to study the origin (γένεσιν) of deductions, but also have the power to produce (ποιεῖν) them" (Aristotle, 1989, A28, 44b, 26, p. 46). In this perspective syllogisms are concerned with problems, when seen as regarding situations in which the conclusion is already given, and the problem is how to find/discover premisses in order to set up a syllogism.[23]

The procedure above, in particular, reduces to a procedure for finding the middle term: during the Middle Age it was called *inventio medii*, that is the discovery of the middle term because it is essentially a procedure for finding the appropriate middle term of a syllogism, given the conclusion. The *Pons Asinorum*, or "bridge of asses", is a diagram embodying Aristotle's *rules of argument-finding* of the *Prior Analytics* I have just quoted. Hamblin (1976, p. 131) contends that "these rules were elaborated by Alexander of Aphrodisias, who speaks of a 'diagram' and may reasonably be credited with having invented it".

Hamblin also reports that the diagram itself seems given by John Philoponus, it is discussed by Averroes, Albert the Great, by the major logicians in the thirteenth and fourteenth centuries, and Bochenski finds it in Thomas Bricot's commentary on the work of George of Brussels. In the fifteenth century it appears in John Dorp's commentary on Buridan, and in Peter Tartaret, who is the first to use the name *pons asinorum* and gives an elaborately drawn diagram with an allegory about an ass (which may be indicates a logical beginner) crossing one of a number of bridges between a major and a minor term. Because of Rabelais the name has become proverbial and it has to be said that it is also sometimes applied to the fifth proposition of the first book of Euclid, concerning the equality of the angles at the base of an isosceles triangle, which is the first difficult theorem in that work

[22] It has to be noted that this abductive need of improving the context seems to reverberate, from an epistemological perspective and perhaps as its main component, the scientist's search for new degrees of discoverability as sources of information in this respect cf. also Hintikka's interrogative model (Hintikka, 2007).

[23] A detailed description of this part of Aristotelian logic is given in cf. (Cellucci, 2013, chapter seven).

and has a figure slightly reminiscent of the logical diagram. Bayle speculates that the name *pons asinorum*, French *pont aux ânes*, arises from a pun between *âne* and *an*, "whether", and refers to the choice between alternative middle terms; he also proposes a link with the fable of Buridan's ass, which could not choose between equally distant heaps of hay. Other authors have connected the doctrine of the pons with Buridan, because of the section *De arte inveniendi medium* of his *Compendium* (Hamblin, 1976, pp. 131–132).

This same procedure for finding the middle term can be seen in the light of the cognitive (and epistemological) *fill-up problem* of abduction, I have quoted above. It is worth to be noted that in dealing with this problem a special role is played by model-based procedures (especially diagrammatic). Indeed, I have first of all to emphasize the role of the method of analysis and of the introduction of new/additional middle terms in Plato's dialectic argumentation,[24] in their relationships with Aristotle's concept of ἀπαγωγή. In the case of the *Meno* dialogue Plato needs to determine whether a certain rectilinear figure could be constructed along the diameter of a circle; to this aim, Socrates establishes a new diagrammatic hypothesis, that incarnates the "leading away" (ἀπαγωγή), which, following Aristotle, opens up the reasoning to an eco-cognitive dimension, which in our case corresponds to an abduction, endowed with its degrees of uncertainty. In a previous article (Magnani, 2016) I also suggested that the function played by diagrams in reasoning was also typical of diorismic/porismic processes and of their creative counterparts, eventually endowed with an objection-refuting role. Levy eloquently says that in Peirce's theorematic reasoning, typical of productive geometrical diagrammatic constructions, "a foreign idea" is introduced, that leads away from previous definitions or known results suggesting new cognitive chances and so an optimization of what I have called in this chapter eco-cognitive situatedness (Levy, 1997, p. 97).[25]

An echo of this reference to the importance of diagrams in analyzing abductive reasoning is indeed still vivid in Peirce, also thanks to the emphasis on the role of icons: "I said, Abduction, or the suggestion of an explanatory theory, is inference through an Icon" (Peirce, 1997, p. 276). It is well known that, for Peirce, signs are icons, indexes, and depending on the nature of the relationship of them to the

[24] An interesting recent revival of these dialectical roots of abduction is represented by the formal treatment and justification of abduction as IBE (inference to the best explanation) in terms of meta-arguments, called by Betz *practical meta-syllogisms* (Betz, 2013). The so-called "dialectical structures" are introduced, which involve the maximization of robustness of one's point of view in an interplay of negotiations.

[25] See also (Campos, 2010, p. 131) and (Marietti, 2010, pp. 153–154). On theorematic and corollarial reasoning in mathematics and on "geometrical constructions interpreted as manipulative abductions" cf. (Magnani, 2009, chapter three, section 3.6.3). It is here very interesting to note that the fifth proposition of the first book of Euclid's *Elements* —I have quoted above when dealing with the procedures for finding the middle term and the *Pons Asinorum*—is also one of Peirce's favorite examples in order to illustrate theorematic diagrammatic reasoning in geometry (Campos, 2010, p. 132).

objects signified. An icon[26] is the middle term of abduction (Trajkovski, 2014), because Peirce aims at stressing the role of *resemblance*, such as in the following case, a syllogism of the second figure [Aristotle insists that all syllogisms are valid (by definition) (Woods, 2014, p. 150), there is no such thing as an invalid syllogism. We know the syllogistic tradition began to relax this requirement quite early on. I am using the term syllogism in this extended sense]:

This woman is pale,
Being with child is accompanied with paleness; therefore
This woman is with child.

where "to be pale" represents the middle term.[27] It *functions* as an icon: a pale women resembles a woman who gave birth, she has some features in common with her, and she looks like her. In general, abduction and icon share the common feature of *creativity*, in so far as an icon can present—through resemblance—*unexpected outcomes*.

On the contrary, index is related to deduction, such as in the case of the middle term "to have milk", which is *taken* as an index, a natural sign of giving birth to a child, see the following related syllogism of the first figure:

This woman has milk,
Having milk means to be with child; therefore
This woman is with child.

In the following example (syllogism of the third figure):

Pittacus is wise,
Pittacus is good; therefore
Wise men are good

"to be Pittacus" is the middle term. Pittacus is, so to speak, a paradigmatic case of wise men, taken by convention, and so represents the foundation for an inductive generalization, not an index, not an icon, but precisely *taken* as a symbol.

## Summary

In this chapter I explained the fundamental aspects of discoverability taking advantage of the Eco-Cognitive model of abduction, fruit of my studies of the last two decades. To this aim I have also taken advantage of a logical formalization

[26] An index instead refers to deduction and a symbol refers to induction, see below.

[27] This and the following examples—reworked in a Peircean perspective—are directly derived from the part of *Prior Analytics* Aristotle devoted to the enthymemes as syllogisms "from likelihoods or signs" (Aristotle, 1989, B29, 70a, 10, p. 136). I think this part represents a kind of "proto-semiotic" treatment of various aspects of reasoning.

which adopts the general concepts of input and output instead of the classical ones of premisses and conclusions. In this light an abductive inference can be seen as a logical process in which input and output fail to hold each other in an appropriate relation, with the solution involving the modification of inputs, not that of outputs. It is thanks to this model that we can see discoverability jeopardized when input and output fail to hold each other in a "good" relation: consequently, the abductive problem at hand cannot be solved. In the case of scientific creative reasoning, which I have described as characterized by a systematic "maximization" of cognition, the relative creative abductive processes—discoverability—is more likely when an "optimization of eco-cognitive situatedness", is reached, certainly busted by the "maximization of changeability" of both input and output, that is a kind of high "information-sensitiveness". I have finally vindicated the role of irrelevance and implausibility in promoting both discoverability and discovery. Plausibility and relevance even if virtuous cognitive heuristics, are not the only criteria that can guide the creation of abductive hypotheses.

## References

Aliseda, A. (1997). *Seeking explanations: Abduction in logic, philosophy of science and artificial intelligence* (Doctoral dissertation, Amsterdam: Institute for Logic, Language and Computation).

Aliseda, A. (2006). *Abductive reasoning. Logical investigations into discovery and explanation.* Heidelberg/Berlin: Springer.

Aristotle. (1989). *Prior analytics.* Translated by R. Smith. Indianapolis/Cambridge: Hackett Publishing Company.

Barés-Gómez, C., & Fontaine, M. (2021). Between sentential and model-based abductions. A dialogical approach. *Logic Journal of the IGPL, 29*(4), 425–446. Special Issue on "Formal Representations in Model-Based Reasoning and Abduction", ed. by A. Nepomuceno, L. Magnani, F. Salguero, C. Barés-Gómez, M. Fontaine.

Bertolotti, T., & Magnani, L. (2014). An epistemological analysis of gossip and gossip-based knowledge. *Synthese, 191*, 4037–4067.

Betz, G. (2013). Justifying inference to the best explanation as a practical meta-syllogisms on dialectical structures. *Synthese, 190*, 3553–3578.

Brandom, R. (2000). *Articulating reason: An introduction to inferentialism.* Cambridge, MA, Edinburgh: Harvard University Press.

Brown, B., & Priest, G. (2004). Chunk and permeate, a paraconsistent inference strategy. Part I: The infinitesimal calculus. *Journal of Philosophical Logic, 33*(4), 379–388.

Cabrera, F. (2017). Can there be a Bayesian explanationism? On the prospects of a productive partnership. *Synthese, 194*(4), 1245–1272.

Cabrera, F. (2020). Does IBE require a "model" of explanation? *British Journal for the Philosophy of Science, 71*(2), 727–750.

Campos, D. G. (2010). The imagination and hypothesis-making in mathematics: A Peircean account. In M. Moore (Ed.), *New essays on Peirce's mathematical philosophy* (pp. 123–145). Chicago: Open Court.

Campos, D. G. (2011). On the distinction between Peirce's abduction and Lipton's inference to the best explanation. *Synthese, 180*(3), 419–442.

Cellucci, C. (2013). *Rethinking logic: Logic in relation to mathematics, evolution, and method.* Heidelberg/Berlin: Springer.

Einstein, A. (2014). Relativity and the problem of space [1952]. In *Ideas and opinions* (pp. 360–377). Translated by S. Bergmann. New York: Crown Publisher.
Estrada-González, L. (2013). Remarks on some general features of abduction. *Journal of Logic and Computation, 232(1)*, 181–197.
Feyerabend, P. (1975). *Against method*. London-New York: Verso.
Fontaine, M., & Barés-Gómez, C. (2019). Conjecturing hypotheses in a dialogical logic for abduction. In D. Gabbay, L. Magnani, W. Park, & P. A (Eds.), *Natural arguments, a tribute to John Woods* (pp. 379–414). London: College Publications.
Gabbay, D. M., & Woods, J. (2005). *The reach of abduction*. Amsterdam: North-Holland.
Gibson, A., & Bruza, P. (2021). Transepistemic abduction: Reasoning across epistemic domains. *Logic Journal of the IGPL 29*(4), 469–482. Special Issue on "Formal Representations in Model-Based Reasoning and Abduction", ed. by A. Nepomuceno, L. Magnani, F. Salguero, Barés-Gómez, M. Fontaine.
Hamblin, C. L. (1976). An improved Pons Asinorum. *Journal of the History of Philosophy, 14(2)*, 131–136.
Hendricks, F. V., & Faye, J. (1999). Abducting explanation. In L. Magnani, N. J. Nersessian, & P. Thagard (Eds.), *Model-based reasoning in scientific discovery* (pp. 271–294). New York: Kluwer Academic/Plenum Publishers.
Hintikka, J. (2007). *Socratic epistemology. Explorations of knowledge-seeking by questioning*. Cambridge: Cambridge University Press.
Kakas, A., Kowalski, R. A., & Toni, F. (1993). Abductive logic programming. *Journal of Logic and Computation, 2*(6), 719–770.
Kowalski, R. A. (1979). *Logic for problem solving*. New York: Elsevier.
Kuipers, T. A. F. (1999). Abduction aiming at empirical progress of even truth approximation leading to a challenge for computational modelling. *Foundations of Science, 4*, 307–323.
Levy, S. H. (1997). Peirce's theoremic/corollarial distinction and the interconnections between mathematics and logic. In N. Houser, D. D. Roberts, & J. Van Evra (Eds.), *Studies in the logic of Charles Sanders Peirce* (pp. 85–110). Bloomington: Indiana University Press.
Livio, M. (2013). *Brilliant blunders: From Darwin to Einstein. Colossal mistakes by great scientists that changed our understanding of life and the universe*. New York, NY: Simon & Schuster.
Magnani, L. (2001). *Abduction, reason, and science. processes of discovery and explanation*. New York: Kluwer Academic/Plenum Publishers.
Magnani, L. (2007). *Morality in a technological world. Knowledge as duty*. Cambridge: Cambridge University Press.
Magnani, L. (2009). *Abductive cognition. The epistemological and eco-cognitive dimensions of hypothetical reasoning*. Heidelberg/Berlin: Springer.
Magnani, L. (2011). *Understanding violence. The intertwining of morality, religion, and violence: A philosophical stance*. Heidelberg/Berlin: Springer.
Magnani, L. (2015a). Naturalizing logic. Errors of reasoning vindicated: Logic reapproaches cognitive science. *Journal of Applied Logic, 13*, 13–36.
Magnani, L. (2015b). The eco-cognitive model of abduction. 'Απαγωγή now: Naturalizing the logic of abduction. *Journal of Applied Logic, 13*, 285–315.
Magnani, L. (2016). The eco-cognitive model of abduction. Irrelevance and implausibility exculpated. *Journal of Applied Logic, 15*, 94–129.
Magnani, L. (2017). *The abductive structure of scientific creativity. An essay on the ecology of cognition*. Cham, Switzerland: Springer.
Magnani, L. (2019). AlphaGo, locked strategies, and eco-cognitive openness. *Philosophies, 4*(1), 8.
Magnani, L. (2021). Abduction as "leading away". Aristotle, Peirce, and the importance of eco-cognitive openness and situatedness. In J. Shook & S. Paavola (Eds.), *Abduction in cognition and action: Logical reasoning, scientific inquiry, and social practice* (pp. 77–105). Cham: Springer.
Marietti, S. (2010). Observing signs. In M. Moore (Ed.), *New Essays on Peirce's mathematical philosophy* (pp. 147–167). Chicago: Open Court.

Mcauliffe, W. H. B. (2015). How did abduction get confused with inference to the best explanation? *Transactions of the Charles S. Peirce Society, 51*(3), 300–319.
McJohn, S. M. (1993). On uberty: Legal reasoning by analogy and Peirce's theory of abduction. *Willamette Law Review, 29*, 191–235.
Meheus, J., Verhoeven, L., Van Dyck, M., & Provijn, D. (2002). Ampliative adaptive logics and the foundation of logic-based approaches to abduction. In L. Magnani, N. J. Nersessian, & C. Pizzi (Eds.), *Logical and computational aspects of model-based reasoning* (pp. 39–71). Dordrecht: Kluwer Academic Publishers.
Park, W. (2015). On classifying abduction. *Journal of Applied Logic, 13*, 215–238.
Peirce, C. S. (1997). *Pragmatism as a principle and method of right thinking. The 1903 Harvard lectures on pragmatism*. Ed. by P. A. Turrisi. (Peirce, C. S., "Lectures on Pragmatism", Cambridge, MA, March 26–May 17, 1903, reprinted in Peirce, 1992–1998, II, pp. 133–241). Albany, NY: State University of New York Press.
Peirce, C. S. (1866–1913). *Collected papers of Charles Sanders Peirce*. Vols. 1–6, C. Hartshorne & P. Weiss (Eds.), vols. 7–8, A. W. Burks (Ed.) (1931–1958). Cambridge, MA: Harvard University Press.
Peirce, C. S. (1992–1998). *The Essential Peirce. Selected Philosophical Writings*. Vol. 1 (1867–1893), ed. by N. Houser & C. Kloesel; vol. 2 (1893–1913) ed. by the Peirce Edition Project. Bloomington: Indiana University Press.
Pietarinen, A. (2020). Abduction and diagrams. *Logic Journal of the IGPL*. Special Issue on "For mal Representations in Model-Based Reasoning and Abduction", ed. by A. Nepomuceno, L. Magnani, F. Salguero, C. Barés-Gómez, M. Fontaine. Online first: https://doi.org/10.1093/jigpal/jzz034
Trajkovski, M. (2014). The origin of semiotic validity—Peirce and Aristotle on reasoning by signs. Presented at the *Charles S. Peirce Centennial Congress 2014*. Lowell, MA. Unpublished manuscript.
van Benthem, J. (2007). Abduction at the interface of logic and philosophy of science. *Theoria, 60(22/3)*, 271–273.
Woods, J. (2007). Ignorance and semantic tableaux: Aliseda on abduction. *Theoria, 60(22/3)*, 305–318.
Woods, J. (2011). Recent developments in abductive logic. *Studies in History and Philosophy of Science, 42(1)*, 240–244. Essay Review of L. Magnani, *Abductive Cognition. The Epistemological and Eco-Cognitive Dimensions of Hypothetical Reasoning*, Springer, Heidelberg/Berlin, 2009.
Woods, J. (2013). *Errors of reasoning. Naturalizing the logic of inference*. London: College Publications.
Woods, J. (2014). *Aristotle's earlier logic*. Second revised edition. Originally published by Hermes Science Publications, Oxford, 2001. London: College Publications.
Woods, J. (2017). Reorienting the logic of abduction. In L. Magnani & T. Bertolotti (Eds.), *Springer handbook of model-based science* (pp. 137–150). Cham: Springer.
Zardini, E. (2014). Context and consequence. An intercontextual substructural logic. *Synthese, 191*, 3473–3500.

# Chapter 2
# Curing Eco-Cognitive Situatedness

## Diagnosticability, Affordances, Abduction

Recent research in the area of the so-called EEEE cognition (extended, embodied, embedded, enacted) has shown that human cognition and its evolutionary dimension can be better understood in terms of their environmental situatedness. This means humans do not retain in their memory rich representations of the environment and its variables, but they actively manipulate it by picking up information and resources upon occasion, already available, or extracted/created and made available: information and resources are not only given but they are actively sought and even manufactured. In this sense, we can consider human cognition as a chance-seeker system. Consequently, in my terminology, chances are not simply information, but they are also "affordances", namely, environmental anchors that allow us to better exploit external resources. Discoverability is of course related to the availability of the appropriate affordances. Abduction is still important: it describes all those human and animal hypothetical inferences[1] that are operated through actions made up of smart manipulations of the environment to both detect new affordances and to create manufactured external objects that offer new affordances/cues. A final section deals with a defense of the inferential view of abduction, affordances, and

[1] Usually the concept of inference is interpreted in logical or psychological terms. On the contrary, as I will clearly explain below in this chapter, I consider the term inference (and so the hypothetical abductive inference) in a Peircean perspective and, consequently, not necessarily concerning rationality. From Peirce's philosophical/semiotic point of view, all thinking is in signs, and signs can be icons, indices or symbols: moreover, all inference is a form of sign activity, where the word sign includes "feeling, image, conception, and other representation" (Peirce, 1866–1913, 5.283), and in Kantian words, all synthetic forms of cognition. In this sense, the term "inference" does not only refer to conscious processes and also deals with cognitive activities involved in model-based and manipulative cognition. This wide meaning attributed to the concept of inference is also related to my eco-cognitive model of abduction, in which, in summary, "eco-cognitive" means that knowing is to be understood in relation to an embodied subject who interacts with his surrounding—that is, who receives and perceives information, but also manipulates his environment whether directly or employing the construction of artifacts and epistemic mediators (Magnani, 2009).



L. Magnani, *Discoverability*, Logic, Argumentation & Reasoning 26,
https://doi.org/10.1007/978-3-030-93329-6_2

cognitive niche construction (based on epistemology and cognitive science), and of its importance for the analysis of discoverability, against the reductionist attitudes due to some interpretations provided by the predictive processing theory, based on the idea of a Bayesian brain.

## 2.1 Various Types of Abduction Depend on the Character of Eco-Cognitive Situatedness

### *2.1.1 "Ecological Validity": Data as Appropriate Affordances that Trigger Abductive Cognition*

Gibson defines "affordance" as what the environment offers, provides, or furnishes. For instance, a chair affords an opportunity for sitting, air breathing, water swimming, stairs climbing, and so on. By cutting across the subjective/objective frontier, affordances refer to the idea of agent-environment mutuality. Gibson did not only provide clear examples, but also a list of definitions (Wells, 2002) that may contribute to generating possible misunderstanding:

1. affordances are opportunities for action;
2. affordances are the values and meanings of things which can be directly perceived;
3. affordances are ecological facts;
4. affordances imply the mutuality of perceiver and environment.

Our problem in this subsection is related to the relationship between abduction and affordances: indeed human and non-human animals can "modify" or "create" affordances by manipulating their cognitive niches[2] so favoring or impeding certain abductive results. Moreover, it is obvious to note that human, biological bodies themselves evolve: and so we can guess that even the more basic and hardwired perceptive affordances available to our ancestors were very different from the

[2] Representational delegations to the external environment that are configured as parts—some of them can play the role of pregnances (see Lexicon of Discoverability, this book)—of *cognitive niches* are those cognitive human actions that transform the natural environment into a cognitive one. Humans have built huge cognitive niches, characterized by informational, cognitive, and, finally, computational processes, as described by the studies in the field of biosciences of evolution by Odling-Smee, Laland, and Feldman (Odling-Smee et al., 2003; Laland & Sterelny, 2006; Laland & Brown, 2006).

present ones.[3] Of course different affordances can also be detected in children, and in the whole realm of animals.

In his research Gibson basically referred to "direct" perception, which does not require the internal inferential mediation or processing by the agent. In this sense affordances express the complementary nature of an organism and its environment through the direct, effortless way in which an organism picks them up (Wells, 2002). It is significant to note that Donald Norman modifies the original Gibsonian notion of affordance also involving mental/internal processing: "I believe that affordances result from the mental interpretation of things, based on our past knowledge and experience applied to our perception of the things about us" (Norman, 1988, p. 14). The same event or place can have different affordances to different organisms but also multiple affordances to the same organism. Following Donald Norman's perspective, affordances suggest a range of *chances*: given the fact that artifacts are complex things and their affordances normally require high-level supporting information, it is more fruitful to study them following this view. To give an example, perceiving the full range of the affordances of a door requires complex information about for example direction of opening or about its particular pull (Scarantino, 2003, pp. 953–954). Among the various chances offered by affordances there are of course the chances which are more or less inclined to offer a significant basis for human discovery, for example in science.

As I have said before, becoming attuned to invariants and disturbances often goes beyond the mere Gibsonian direct perception, and higher representational and mental processes of thinking/learning have to be involved. This means that for example in designing an artifact to the aim of properly and usefully exhibiting its full range of affordances we have to clearly distinguish among two levels: (1) the construction of the utility of the object and (2) the delineation of the possible (and correct) perceptual information/cues that define the available affordances of the artifact. They can be more or less easily be undertaken by the user/agent (Gaver, 1991; Warren, 1995; McGrenere & Ho, 2000): "In general, when the apparent affordances of an artifact match its intended use, the artifact is easy to operate. When apparent affordances suggest different actions than those for which the object is designed, errors are common and signs are necessary" (Gaver, 1991, p. 80). In this last case affordances are apparent because they are simply "not seen". In this sense information arbitrates the perceivability of affordances, and we know that available information often goes beyond what it can be provided by direct perception but instead involves higher cognitive endowments.

[3] The term "wired" can be easily misunderstood. Generally speaking, I accept the distinction between cognitive aspects that are "hardwired" and those which are simply "pre-wired". By the former term, I refer to those aspects of cognition which are fixed in advance and not modifiable. Conversely, the latter term refers to those abilities that are built-in prior the experience, but that are modifiable in later individual development and through the process of attunement to relevant environmental cues: the importance of development, and its relation with plasticity is captured thanks to the above distinction. Not all aspects of cognition are pre-determined by genes and hardwired components. Cf. also (Barrett & Kurzban, 2006).

A true situation of distributed cognition is occurring in the case of online thinking, like in the case of manipulative abduction[4] and in other less expert and less creative cases, where the resources are not merely inner (neurally-specified) and embodied but hybridly intertwined with the environment: in this case we face with an abductive/adaptive process produced in the dynamical inner/outer coupling where internal elements are "directly *causally locked onto* the contributing external elements" (Wheeler, 2004, p. 705).

Following Brunswik's theory an organism is not able to perceive distal stimuli directly but instead must *infer* what is going on in its ecological niche from the cues available (cues provided by proximal stimuli). The success (Brunswik says the *ecological validity*) of this "vicarious" inference[5] is of course jeopardized by the constitutive incompleteness (and unreliability, ambiguity, and equivocality) of the cues available and by their highly variable diagnostic character: Brunswik, implicitly professing an abductive attitude worthy of Peirce, says: "[...] ordinarily organisms must behave as if in a semierratic ecology" (Brunswik, 1955, p. 209), given the intrinsic "ambiguity in the causal texture of the environment" (Brunswik, 1943, p. 255). In this sense he adds that the cues and the mediating inference are both "probabilistic", like in the case of abduction where it is always the case that: "Both the object-cue and the means-end relationship are relations between probable partial causes and probable causal effects" (Brunswik, 1943, p. 255).

In this light the Brunswikian concept of *ecological validity* can be seen in terms of abductive plausibility of the inference at play, given the available data/cues: in other words, ecological validity coincides with discoverablity/diagnosticability. The quality of the inferential abductive performance measures the degree of adaptation between an organism's behavior and the structure of the environment, i.e. the fitness of the behavior based on the particular adopted inference. When the cues are object of simple and immediate perceptual appraisal the situation reflects what I have illustrated in the case of the so-called "visual abduction" (Magnani, 2009, chapter two).[6] On the contrary, in the other cases, organisms more or less appropriately inferentially abduce a "hypothesis/judgment" about distal structure of the environment. It is worthy to quote again Gibson's intuition, which can better be grasped in this perspective: "Perceiving is the simplest and best kind of knowing.

[4] The concept of *manipulative abduction*—which also takes into account the external dimension of abductive reasoning in an eco-cognitive perspective—captures, to make a clear example, a large part of scientific thinking where the role of action and of external models (for example diagrams and artefacts) and devices is central, and where the features of this action are implicit and hard to be elicited. Action can provide otherwise unavailable information that enables the agent to solve problems by starting and by performing a suitable abductive process of generation and/or selection of hypotheses. Manipulative abduction happens when we are "thinking through doing" and not only, in a pragmatic sense, about doing (Magnani, 2009, chapter one).

[5] Again, here the word inference has to be intended in the expanded semiotic Peircean sense, i.e. the term must not collapse in the restricted logical meaning.

[6] In this last case we could say there is a one-to-one mapping between proximal and distal structure (Vicente, 2003, p. 261).

But there are other kinds, of which three were suggested. Knowing by means of instruments extends perceiving into the realm of the very distant and the very small; [...] Knowing by means of language makes knowing explicit instead of tacit" (Gibson, 1979, p. 263). An example can be a forecast—usually probabilistic—about the behavior of wind-based on the current wind speed measured at a ground station as illustrated in a computer screen—the "cue". It is important to say that in this last case the proximal perception affords the day-after action of wearing clothes suitable to the weather.

I have said above that in the mediating inferential processes also cognitive delegations to the environment (for example to automated artifacts) can also play a role, for example in facilitating reliable action/decisions.[7] In both human and non-human animals artifacts can reduce the uncertainty of the relationship between organism and environment. Even in these cases, for example—in the case of humans—when technology is directly designed to respond with greater precision than people can do, a person's abductive judgment can still fail to correspond with the distal environment. Research based on Brunswikian tradition has emphasized the essential ecological character of the cognitive engineering enterprise in the framework of systems composed of the interaction among humans, mediating technologies, and tasks environments. Many results have shown in various interesting ways how many technological devices help humans to fulfill their adaptation to the environment by enhancing hypothesis generation/judgment and, consequently, decision-making. Often the technology itself fails to provide the correct judgment about a given situation, in other cases the gap in the proximal/distal relationship is embedded in the interaction with the user (Kirlik, 2006b).

For example (Byrne et al., 2006) study the airline pilot's performance in landing and taxiing to the gate in foggy conditions by showing, thanks to an ecological analysis, how the cockpit artifacts (and their interfaces which provide proximal cues/chances for action) affect the overall cognitive performance. Often the problem is related to the fact that technologies exhibit a *discrete* ecology that does not sufficiently involve the approximation and convergence performed by *continuous* ecologies of natural environmental structures. This fact for example explains why it has been shown that the current design of cockpit automation leaves pilots less supported in special uncertain (and more challenging)—unsafe—situations: control systems proximal to pilots are discrete, whereas the behavior of the distal controlled system—the aircraft—is continuous (Degani et al., 2006). Finally, in the mediating abductive inferences occurring through artifacts, time (and the so-called "time-stress" and "time-pressure" effects and their relationship to knowledge deficits and task simplification) is still a key element that strongly affects information incompleteness and the degree of success of various heuristic schemes.

Interpreting perception and other forms of cognition as ways of seeking salient information that involves extra-neural enactive processes related to bodies and social environment rejoins what I have illustrated in the previous chapter about

[7] The case of computer interfaces and time pressure is treated in (Adelman et al., 2006).

cognitive *action* and its "situatedness". I have said it is a way of getting more sensory data, compensating for their equivocality, and obtaining cognitive feedback, and/or a way of manipulating them, and also of exploiting *cognitive delegations* to the environment and to artifacts. In this sense, brains do not need to form rich internal models of the environment and so they do not store "[...] all knowledge about the world in explicit form and do not hold comprehensive explicit models or representations of the environment. Rather, it has been argued [...] our brains contain recipes for seeking out information—often rather trivially by deciding where in the environment to look" in the interplay between internal and external (delegated) representation (Adolphs, 2006, p. 32).

### *2.1.2 Affordances, Diagnosticability, and Creative Abduction: The Plasticity of Environmental Situatedness*

I have quoted above Gibson's conviction that "The hypothesis that things have affordances, and that we perceive or learn to perceive them, is very promising, radical, but not yet elaborated" (Gibson, 1979, p. 403). Let us deepen this issue: we can say that the fact that a chair affords sitting means we can perceive some clues (robustness, rigidity, flatness) from which a person can easily say "I can sit down". Now, suppose the same person has another object $O$. In this case, the person can only perceive its flatness. He/she does not know if it is rigid and robust, for instance. Anyway, he/she decides to sit down on it and he/she does that successfully. Again, we are faced with the problem of direct and indirect visual perception. It is thanks to the effect of action that we can detect and stabilize the new affordances.

Now, my point is that we should distinguish between the two cases: in the first one, the cues we come up with (flatness, robustness, rigidity) are *highly diagnostic* to know whether or not we can sit down on it, whereas in the second case we eventually decide to sit down, but we do not have any precise clue about it. How many things are there that are flat, but one cannot sit down on? A nail head is flat, but it is not useful for sitting. This example further clarifies two important elements: firstly, finding/constructing affordances certainly deals with a (semiotic) inferential activity (cf. Windsor, 2004); secondly, it stresses the relationship between an affordance and the information that specifies it that only arises in the *eco-cognitive interaction* between environment and organisms. In this last case, the information is reached through a simple action, in other cases through action and complex manipulations.

The term "highly diagnostic" explicitly refers to the abductive framework. In chapter one of my book on abduction (Magnani, 2009), I have said that abduction is a process of *inferring* certain facts and/or laws and hypotheses that render some sentences plausible, that *explain* or *discover* some (eventually new) phenomenon or observation. The distinction between theoretical and manipulative abduction extends the application of that concept beyond the internal dimension. I repeatedly said that from Peirce's philosophical point of view, all thinking is in signs, and signs can be icons, indices, or symbols: moreover, all inference is a form of sign activity, where the word sign includes "feeling, image, conception, and other representation" (Peirce, 1866–1913, 5.283), and, in Kantian words, all synthetic forms of cognition. That is, a considerable part of the thinking activity is "model-based" and consequently non-sentential. Of course model-based reasoning acquires its peculiar creative relevance when embedded in abductive processes, so that we can individuate a *model-based abduction*. In the case of diagnostic reasoning in medicine, a physician detects various symptoms (that are signs or clues) in a multimodal way, for instance, cough, chest pain, and fever, *then* he/she may infer that it is a case of pneumonia.

The original Gibsonian notion of affordance especially deals with those situations in which the "perceptual" signs and clues we can detect prompt or suggest a certain action rather than others.[8] They are already available and belong to the normality of the adaptation of an organism to a given ecological niche. Nevertheless, if we acknowledge that environments and organisms' instinctual and cognitive plastic endowments change, we may argue that affordances can be related to the variable (degree of) *abducibility* of a configuration of signs: *a chair affords sitting* in the sense that the action of sitting is a result of a sign activity in which we perceive some physical properties (flatness, rigidity, etc.), and therefore we can ordinarily "infer" (in Peircean sense) that a possible way to cope with a chair is sitting on it. So to say, in most cases, it is a spontaneous abduction to find affordances because this chance is already present in the perceptual and cognitive endowments of human and non-human animals.

I maintain that describing affordances that way may clarify some puzzling themes proposed by Gibson, especially the claim concerning the fact that we directly perceive affordances and that the value and meaning of a thing is clear at first glance: organisms have at their disposal a standard endowment of affordances (for instance through their hardwired sensory system), but at the same time they can extend and modify the scope of what can afford them through the suitable cognitive abductive skills. I also stressed the important fact that environments change but also perceptive capacities enriched through new or higher-level cognitive skills change, those capacities that go beyond the ones granted by the merely instinctual levels: if affordances are usually stabilized this does not mean they cannot be modified and changed and that new ones can be formed.[9]

---

[8] As I have already indicated, in the original Gibsonian view the notion of affordance is mainly referred to proximal and immediate perceptual chances, which are merely "picked up" by a stationary or moving perceiver. I maintain that perceiving affordances also involves evolutionary changes and the role of sophisticated and plastic cognitive capacities (further details below in this subsection).

[9] Thom nicely says that "Nature (φύσις) is present in the behavior of inanimate beings. But the animate being is able to exploit natural regularities in order to stabilize connections that would be accidental, not generic, in the inanimate world" (Thom, 1988, p. 217). On Thom's catastrophe theory and its consequences for the concept of abduction cf. (Magnani, 2009, chapter eight).

First of all, affordances appear durable in human and animal behavior, like kinds of habits, as Peirce would say (Peirce, 1866–1913, 2.170). For instance, that a chair affords sitting is a fair example of what I am talking about. This deals with what we may call *stabilized affordances*. That is affordances that we have experienced as highly successful. Once evolutionarily formed, or created/discovered through cognition, they are stored in embodied or explicit cognitive libraries and retrieved upon occasion. Not only, they can be a suitable source of new chances, through analogy. We may have very different objects that equally afford sitting. For instance, a chair has four legs, a back, and it also stands on its own. The affordances exhibited by a traditional chair may be an analogical source and transferred to a different new artifact that presents the affordance of a chair for sitting down (and that to some extent can still be described as a chair). Consider, for instance, the variety of objects that afford sitting without having four legs or even a back. Let us consider a stool: it does not have even a back or, in some cases, it has only one leg or just a pedestal, but it affords sitting as well as a chair.

Second, affordances are also subjected to changes and modifications. Some of them can disappear because new configurations of the cognitive environmental niche (for example new artifacts) are invented with more powerful offered affordances. Consider, for instance, the case of blackboards. Progressively, teachers and instructors have partly replaced them with new artifacts which exhibit affordances brought about by various tools, for example, slide presentations. In some cases, the affordances of blackboards have been totally re-directed or re-used for more specific purposes and actions. For instance, one may say that a logical theorem is still easier to be explained and understood by using a blackboard, because of its affordances that give a temporal, sequential, and at the same time global perceptual depiction to the matter.

Of course—in the case of humans—objects can afford different persons in different ways. This is also the case of experts: they take advantage of their advanced knowledge within a specific domain to detect signs and clues that ordinary people cannot detect. For instance, a patient affected by pneumonia affords a physician in a completely different way compared with that of any other uncultured person. Being abductive, the process of perceiving affordances mostly relies on a continuous activity of hypothesizing which is cognition-related. That $A$ affords $B$ to $C$ can be also considered from a semiotic perspective as follows: $A$ signifies $B$ to $C$. $A$ is a sign, $B$ the object signified, and $C$ the interpretant. Having cognitive skills (for example knowledge contents and inferential capacities but also suitable pre-wired sensory endowments) about a certain domain enables the interpretant to perform certain abductive inferences from signs (namely, perceiving affordances) that are not available to those who do not possess those apparatuses. To ordinarily people a cough or chest pain are not diagnostic, because they do not know what the symptoms of pneumonia or other diseases related to cough and chest pain are. Thus, they cannot make any abductive inference of this kind and so perform subsequent appropriate medical actions.

## 2.2 Discoverability and Diagnosticability through Affordance Creation

Consider, for instance, a huge stone and a chair. Indeed, both these objects afford sitting. The difference is that in the case of a stone affordances are simply already, so to say, *given*: we find a stone and we ordinarily "infer" (in Peircean sense) that it can be useful for sitting. In contrast, those of a chair are somehow *manufactured* from scratch. In the case of a chair, we have entirely created an object that displays a set of affordances. This process of building affordances can be made clearer by taking advantage of the abductive framework we have introduced above.

That an object affords a certain action means that it embeds those *signs* from which we "infer"—through various cognitive endowments, both instinctual and learned—that we can interact with it in a certain way. As already said, in the case of a stone, humans exploit a pre-existing configuration or structure of meaningful sign data that are in some sense already established in the interaction organism/environment shaped by past evolution of the human body (and in part "material cultural" evolution, for example when hominids exploited a stone/chair to sit down in front of a rudimentary altar). In the case of a chair, this configuration is invented. If this perspective is correct, we may argue that building "artificial" affordances means configuring signs in an external environment expressly to trigger new proper inferences of affordability. In doing this, we perform smart manipulations and actions that—I conjecture—can produce new (and sometimes "unexpected") affordances. Accordingly, creating affordances is at the same time making new ways of inferring them feasible: a process that is strictly related to the enhancement of diagnosticability and discoverability.[10]

### *2.2.1 Latent Constraints*

The organism's "end-stage" of being afforded, in the Brunswikian interplay proximal/distal, not only involves perception, but also a more complicated *inferential processing* which in turn occurs through either

1. a mere *internal* cognitive processing (expressed by model-based abduction and, in the case of human beings, also by sentential abduction, cf. (Magnani, 2009, chapter one)
   or
2. a (cognitive) composite *action* with the aim of getting more sensory data, compensating for their equivocality, and obtaining cognitive feedback (confirmatory and disconfirmatory), and/or with the aim of manipulating them (manipulative

[10] On this note, see (Magnani & Bardone, 2005) about the role of abduction in designing computer interfaces.

abduction), also exploiting *cognitive delegations* to the environment and to *artifacts*.
[Indeed these high-level inferential processes affect perception in various ways, like it is shown by the evidence coming from the studies illustrated above].

I have consequently said that affordances emerge through perceptions that in humans are basically semi-encapsulated (that is also involving a top-down "interpretation" of data), and also affected by the two processes above and that grant the subsequent (pragmatic) *action*.

Consequently, two kinds of "actions", cognitive and pragmatic (performatory) are at play. Kirlik (2006a), also resorting to a distinction between pragmatic and epistemic action offered by Kirsh and Maglio (1994),[11] offers an analysis of the problem that is extremely interesting to further clarify the dichotomy. In Kirlik's words, the first type of action I have just indicated plays an "epistemic" or "knowledge-granting" role and the second a "performatory role in the execution of interactive tasks" (Kirlik, 2006a, p. 214). It is well-known that traditionally, cognitive scientists have seen action systems in a purely performatory/pragmatic light, thus lacking any immediate "cognitive" features other than the ability to execute commands.

Recent research, taking an explicit ecological approach to the analysis and design of human-machine systems (Kirlik, 1998), has shown how expert performers use action in everyday life to create an *external* model of task dynamics that can be used in lieu of an internal model: a child shaking a birthday present to guess its contents is dithering, a common human response to perceptually impoverished conditions. Not only a way for moving the world to desirable states, action performs an epistemic and not merely performatory role: people structure their worlds to simplify cognitive tasks but also in the presence of incomplete information or faced with a diminished capacity to act upon the world when they have fewer opportunities to gain knowledge. Epistemic action can also be described as resulting from the exploitation of latent constraint in the human-environment system. This additional constraint grants additional information and so enhances diagnosticability: in the example of the child shaking a birthday present she is taking action that will cause variables relating to the contents of the package to covary with perceptible acoustic and kinesthetic variables. Epistemic actions result from exploiting *latent constraints* in the human-environment system as chances for further inferences. "Prior to shaking, there is no active constraint between these hidden and overt variables causing them to carry information about each other". Similarly, "[...] one must put a rental car 'through its paces' because the constraints active during normal, more reserved driving do not produce the perceptual information necessary to specify the dynamics of the automobile when driven under more forceful conditions" (Kirlik, 1998, p. 24) . In this light (Powers, 1973) studied behavior considering it as a *control*

[11] Kirsh describes everyday situations (e. g. grocery bagging, salad preparation), where people adopt actions to simplify choice, perception, and reduce the amount of internal computation through the use of suitable cognitive delegations to the environment and to artifacts Kirsh, 1995.

*of perception* and not only as controlled by perception. (Flach & Warren, 1995) used the term "active psychophysics" to illustrate that "[...] the link between perception and action [...] must be viewed as a dynamic coupling in which stimulation will be determined as a result of subject actions. It is not simply a two-way street, but a circle" (p. 202).

Moreover, a very interesting experiment is reported concerning short-order cooking at a restaurant grill in Atlanta: the example shows how cooks with various levels of expertise use external models in the dynamic structure of the grill surface to get new information otherwise inaccessible.

In Brunswikian terms, some variable values are proximal because they can be perceived and others cannot because they are distal. This distinction, Kirlik observes, is relative:

> A particular variable (e.g. the velocity of an automobile) could be described as proximal if the purpose of a study was to understand how a driver might use velocity to infer less readily available information relating to a vehicular control task (e.g. whether he or she can take a given turn safely). In other cases, velocity could be considered a distal variable if the goal of the effort was to understand how velocity itself was estimated on the basis of even more immediate perceptual information (e.g. optical flow patterns, speedometer readings, etc.) (Kirlik, 2006a, p. 216).

Of course from the perspective of action, the values of proximal variables can directly be manipulated. On the contrary distal variables can be changed (and so inferred) only by manipulating proximal variables. On this basis Kirlik illustrates a generalized framework that provides an ontology for describing an environment of a performer who is undertaking an interactive activity. Variables exhibited to an agent in a world of surfaces, objects, events, and artifacts present various values:

1. Type I [PP.PA]: variables are proximal from the perspective of both perception and action. "The location of the book you are now reading can most likely be represented by variables of this type: most of us can directly manipulate the values of these variables by hand and arm movements, and we can also perceive the location of the book directly. One can think of these variables as representing Gibson's directly perceptible affordances" (Kirlik, 2006a, p. 216).
2. Type II [PP.DA]: variables proximal for perception but distal for action. Variables are directly available to perception but we can change their values by manipulating proximal variables that cause changes in the values of the distal action variable. You feel it is cold in the room but you need to manipulate the thermostat to change the temperature and consequently your feeling, but you cannot directly change the temperature.
3. Type III [DP.PA]: variables distal for perception and proximal for action. It is the case that many of my actions through for instance an artifact can change the values of the environment, which I cannot directly perceive. A common outcome of the manipulation of many artifacts is that they for example have moral consequences, perhaps even unexpected ones, for human beings and objects very distant and remote from us, which might last forever. "When posing for a photograph I change the location of my image in the viewfinder of a camera

without being able to perceive how it has been changed" (Kirlik, 2006a, pp. 217–218).[12]

4. Type IV [PP.PA]: variables that are distal from both perception and action, very common in the case of technology. "We infer the values of these variables from interface displays, and we change the value of these variables by using interface controls" (Kirlik, 2006a, p. 218).

In the framework above, which stresses the importance of inner-outer interaction in proximal-distal dynamics, it is very easy to interpret artifacts built by humans as ways of adaptation that, through the construction of suitable cognitive feasible niches, mediate and augment our interaction with the distal world. They aim at enhancing intellectual functioning by offering suitable differentiated affordances/proximal-cues, which are easier to perceptually detect and present new opportunities for action. This strategy, which modifies previous available human "cognitive ecology", can offload some cognitive demands to the world through a better exploitation of artifacts.[13]

### *2.2.2 Creating Chances through Manipulating Artifacts and External Representations*

It is now clear that the history of culture, artifacts, and various tools can be viewed as a continuous process of building and crafting new affordances upon pre-existing ones or even from scratch. From cave art to modern computers, there has been a coevolution between humans and the environment they have built and they live in. Indeed, what a computer can afford embraces an amazing variety of opportunities and chances comparing with the ones exhibited by other tools and devices. More

[12] In (Magnani, 2007) I have illustrated a related issue: many artifacts play the role of "moral mediators". This happens when macroscopic and growing phenomenon of global moral consequences and collective responsibilities result from the "invisible hand" of manipulations of artifacts and interactions among agents at a local level, like for example in the case of the effect of the Internet on privacy or of the derivatives in the international economical crisis: for example my manipulations on the net can affect the privacy of people with effects that I cannot perceive.

[13] Kirlik (2007) describes various interesting examples of the use of artifacts to enhance cognitive skills in technological disciplines like architecture and aeronautics, but also in expert everyday performances like cooking. A negative example of proximal-distal dynamics has to be noted in many contemporary artifact devices (for example in cars) which offer signs that mainly appeal to the visual sense. This involves learning to answer to visual rather than to auditory, olfactory—for example motor smell is extinct in new cars—and tactile proximal cues so loosing out on a lot of important body-mind sensorimotor resources (Bauters, 2007). We can say, in Peircean terms, that in these cases the semiotic ground active in perception and in which the actions can emerge seems poor, allowing bad and unsatisfactory habits to be afforded, promoted, and stabilized (Vandi, 2007): indeed Peirce maintains that habits are "[. . . ] a general law of action, such that on a certain general kind of occasion a man will be more or less apt to act in a certain general way" (Peirce, 1866–1913, 2.148).

precisely, a computer as a Practical Universal Turing Machine (cf. Turing, 1992 and Magnani, 2021b) can mimetically reproduce a considerable part of the most complex operations that the human brain-mind systems carry out (Magnani, 2006). For instance, computers even result in many respects more powerful than humans in memory capacity and in some aspects of mathematical reasoning. From a semiotic perspective, computers bring into existence new artifacts that present "signs" (in Peircean sense) for exploring, expanding, and manipulating our own brain cognitive processing (and so they contribute to "extend the mind beyond the brain"), that is, offering and creating new affordances. As just argued above, building affordances deals with a semiotic activity, mainly abductive, in which signs are appropriately scattered all around in order to prompt (or suggest) a certain interaction rather than others.

To clarify this point, consider the simple diagrammatic demonstration (we all have learned at the middle school) that the sum of the internal angles of any triangle is 180° (I have illustrated this case in (Magnani, 2009, chapter three) as a prototypical case of manipulative abduction. This is also an example of construction of affordances taken from the field of elementary geometry that (1) in case of current children, help arrive at the conclusion of the proof, and (2) in the case of the ancient pre-Euclidean geometers, provided the suitable discoverability able to grant new geometrical results. In this case, a simple manipulation of a suitably externally depicted triangle gives rise—through constructions—to other external configurations that uncover new visual affordances. The process occurs through construction and modification of the initial external diagram, so as we can easily arrive at the generalized result that the sum of the internal angles of any triangle is 180°. The process happens in a distributed interplay between a continuous externalization through cognitive acts, its manipulation, and re-internalization that acknowledges what has been discovered outside, picking up the result and re-internalizing it. In the action new affordances arise and lead to the result whereby, the sum of the internal angle is found.

From the epistemological point of view this example is a typical example of that manipulative abduction I have quoted above. Reframed in terms of affordances this is a cognitive manipulation (entirely abductive, the aim is to find a incontrovertible geometrical hypothesis) in which an agent, when faced with merely "internal" representational geometrical "thoughts", from which it is hard or impossible to extract new meaningful features of a concept, organizes epistemic action that structures the environment in such a way that unearths new affordances as chances which favor new outcomes. As already mentioned, the detection of affordances is hypothesis-driven: it is not said that just anybody can detect it. In the example of the triangle, only a person that has been taught about geometry can infer the affordances within the manipulated construction built upon the original triangle. In this sense, "artificial" affordances are intimately connected to culture and knowledge available in specific human cognitive niches but also to their availability to the suitable individuals involved in the epistemic process.

We can say that in a given geometrical diagram various affordances still exist regardless of correct interpretation or even perception by the agent: indeed the diagram embeds geometrical knowledge that potentially means something to someone. To correctly grasp the affordances

1. the agent has to appreciably "know" geometry,
2. the particularly expert agent—through suitable cognitive manipulations in the internal/external representations interplay—can also extract/find/discover "new" affordances like chances offered to possibly increase knowledge.

The diagram offers nested affordances:[14]

1. it is a simple picture, which almost everybody (and many animals) can see and understand as a perceptual picture that affords potential colors and shapes, depending on the perceptive hardwired endowments of the organism in front of it—for example expert and uncultured persons but also cats—(strict Gibsonian case);
2. it is a picture which can be seen and understood as a geometrical diagram with all its technical properties and features (in this case higher cognitive endowments are needed in the organism at stake: a human being, of course, other animals do not understand geometry),
3. it is an artifact which can offer—through even higher creative cognitive manipulations—new affordances to be picked up and possibly inserted in the available encyclopedia of geometrical knowledge. Imagine a child who must "demonstrate" a theorem of the elementary geometry (the sum of the internal angles of a triangle). The child does not have to demonstrate this theorem for the first time, in the sense that it has already been discovered (demonstrated) historically and reported in every manual of Euclidean geometry. However, excluding the case when he passively repeats by rote, he can achieve this demonstration by using the series of elementary extracted affordances illustrated above, based on the appropriate use of basic geometric concepts which are already available to him. We can also say that the child used a heuristic, that is an advanced procedure of discovery. And this heuristic, naturally, considered from the viewpoint of an already available manual of geometry (as an abstract and static system of knowledge) is a true "demonstration", and obviously does not lead to a discovery. Rather, it is a kind of re-discovery. Also from the viewpoint of the child-subject, it is a re-discovery, because he re-discovers a property that is already given at the beginning). Instead, the inferences that were employed at the moment of the first historical discovery (maybe Greek) of that property

[14] Turner (2005) clearly acknowledges the character of affordances as nested. They are seen as "complex" affordances expressed by "significances" delegated to the natural and artificial environment—in the perspective of an interesting semiotic and Heideggerian approach to distributed cognition. On the role of the "extra-bodily world"—an artifact, for example—that becomes poised itself to the user not just as a problem-space but also as a problem-solving resource via "inhabited interaction", which grants fluent acts on ways that simplify or transform the problems, cf. (Clark, 2008, p. 10). Clark usefully stresses that in case of the inhabited interaction the body has become "transparent equipment" and tools lead to the alterations of the local sense of embodiment: a true incorporation rather than mere use is at play, like Heidegger already contended.

of the triangles (and the assessment of the respective theorem) established a kind of creative achievement (as I have said, a creative manipulative abduction). Moreover, both types of reasoning are also mainly model-based, considering the fact they are performed with "hybrid" forms of representation, including considerable non-verbal devices (like the geometrical diagrams).[15]

The case indicated in item (1) also explains why an affordance can be grasped at the same time by animals, infants, and adult human beings, despite their cognitive differences: all of them can perceive "the brink of a cliff as *fall-off-able* according to a common perceptual process" (Scarantino, 2003, p. 960), even if they have different perceptual endowments. "This is much the same as we would describe a piano as having an affordance of music playability. Nested within this affordance, the piano keys have the affordance of depressability" (McGrenere & Ho, 2000, p. 340). We can add that the piano also offers the chance—discoverability—in the cognitive interplay artifact/agent, of providing new affordances of *new* good melodies, not previously invented in a merely internal way, in the agent's mind, but found over-there, in the hybrid interplay with the artifact. Of course both artifacts, the diagram and the piano, offer in themselves various constrained conditions for affordances, depending on their properties, quality, materiality, and design, and of course various degrees of affordances. From Kirlik's perspective in terms of variables and proximal/distal distinctions the example above can be illustrated in this way: the agent builds a diagram in which he/she can work on the surface by taking advantage of the constraints that guarantee that latent variables intrinsic to the materiality of the artifact at hand "to take care of themselves, so to speak" (Kirlik, 2006a, p. 221).

The need of having a rich internal model of the depicted triangle is weakened because various aspects are discharged from the agent and delegated to the external representation, which offers a proximal perceptual and a manipulative environment containing all the resources needed to effectively perform the creative task of finding new properties of a triangle. The result is immediately perceived and consequently can be picked up and internally re-represented. The diagram itself is a model in the dynamics of the geometrical reasoning and its manipulation expresses an example of situated cognitive action completely intertwined with perception.

From a semiotic perspective, at first glance, we do not have sufficient cognitive capacities to internally infer what the sum of internal angles is. Manipulating the externalized configuration (the external diagram of a triangle) we come up with a new configuration of perceptual signs that adds properties not contained neither in the initial internal representation of a triangle nor in the initial external depiction.

[15] Of course the agent can modify and change in a more or less creative way the features of the artifacts just to make the available affordances visible or more exploitable, or to build new ones that are offered as options. This is for example the case of a user who customizes a computational interface making an alias for a long command string. He/she can gain easier use of the tool by simply invoking a single key or by the simultaneous pressing of multiple keys instead of writing a long string of characters (McGrenere & Ho, 2000).

This new configuration of signs gives birth to a new set of affordances that allow us to find the solution. In a Euclidean sense it is a way of "demonstrating" a new theorem. This example furnishes an epistemological example of the nature of the cognitive interplay between internal neuronal semiotic configurations that permit the representational thought about a triangle (together with the help of various embodied "cognitive" kinesthetic abilities) and external representations: also for Peirce, more than a century before the new ideas derived from the field of distributed reasoning, the two aspects are pragmatically intertwined.

Indeed, the human possibility of creating affordances is constrained in this interplay. More precisely, this possibility depends on:

- the specific cognitive delegation to externalities suitably artificially modified, and
- the particular intrinsic constraints of the manipulated materiality at play, which usefully provide to us new—so to say—"insights".

The first aspect is closely related to the idea of humans as chance-seekers illustrated above. As extensively argued throughout this chapter, securing meaning to the environment relies on smart manipulations that are always hypothetical and therefore withdrawable. That is, it depends on our ability to exploit latent chances. Regarding the second aspect, the environment has constraints of its own. When designing new affordances, we do not work on a blank slate. Conversely, we are in a network of pre-existing ones that constrain our possibilities. On one hand, the possibility of uncovering new latent chances depends on human agency. On the other hand, this agency is mediated/constrained by what the environment offers in terms of pre-existing affordances and its potentiality to provide new ones, as maintained.

This "hypothetical" status of affordances reminds us that it is not necessarily the case that just anybody can detect it. Affordances are a mere potentiality for organisms. First of all perceiving affordances results from an abductive activity in which we infer possible ways to cope with an object from the signs and cues available to us. Some of them are more or less unchanging and in some cases they are neurally hardwired or pre-wired (pre-specified) in the perceptual system—they are "invariants", to exploit a term derived from physics, also used by Gibson: this is the case of affordances that have a high cognitive valence.[16] We said the differences that we can appreciate are mostly *inter-species*—so to speak. In the high-level cognitive performance on triangles there is something different, since *intra-species* differences seem to be strongly involved. For instance, only a person that has been taught about geometry can infer (and so "perceive") the affordances "inside" the new manipulated construction built on the original triangle I have illustrated above. This concerns the problem of "expertise" I already noted. First of all, artificial affordances are intimately connected to culture and the social settings

[16] Perceiving the affordances of a chair is indeed rooted and "stabilized" in our cultural evolution because for human beings it is easy—and *possible*, given our cognitive-biological configuration—to gain the corresponding cognitive ability as a "current" and "reliable" ability (Scarantino, 2003, p. 959).

in which they arise and the suitable availability of knowledge of the individual(s) in question. Secondly, affordances deal with learning. Perceiving some affordances like those of a triangle is not a *built-in* activity, indeed, once manufactured, they can be learned and taught. Of course, acknowledging this last fact places much more emphasis upon the dynamic character of affordances in organisms' plastic cognitive life, beyond their evolutionary character.

Hence, the capacity to perform smart manipulations is concerned with the activity of creating external representations. Humans are continuously engaged in cognitive mimetic and creative processes (cf. Donald, 2001 and Magnani, 2009, chapter three) in which they represent thoughts, private ideas, solutions, into external structures and suitably constructed artifacts. In doing so they create external counterparts of some internal, already available propositional and model-based representations, suitably stored in their brains. In some cases, these external counterparts, which initially just simply mirror ideas or thoughts already present in the mind, later on, can be creatively used to find room for finding new concepts and new ways of inferring that cannot be exhibited by the mere "internal" representation alone (Magnani, 2006). In building these external representations (that—I repeat—can be held as merely mimetic, but that can become "creative"), people manipulate the world in such a way that new cognitive *chances* are uncovered, and so discoverability or diagnosticability enhanced. In doing this, new affordances are thus manufactured and made "socially" available. More generally, we can reiterate, also from this perspective, that abduction also deals with the continuous activity of manipulating the environment in order to create new chances and opportunities to action, namely, affordances. Recent interesting research on the role of models, deriving from 5-year empirical ethnographic study of two systems biology labs and their collaborations with experimental biologists, has provided the analysis of the so-called "adaptive problem solving" (MacLeod & Nersessian, 2016).[17]

## 2.3 Curing Eco-Cognitive Situatedness to Favor Abductive Cognition

An interesting application of the relationship between affordances, diagnosticability, and abduction I am proposing in this chapter is given by (Abe, 2009). In order to support dementia persons' lives, a model-based approach grounded in the concepts

[17] On the role of models to address just some sort of epistemic opacity in biological research that takes advantage of computational visualizations cf. (Stuart & Nersessian, 2018). On the exceptional role played by models in scientific research and in many other disciplines cf. the handbook of model-based science I coedited (Magnani & Bertolotti, 2017). On the crucial role of the propositional imagination in the construction and development of models for the purpose of generating hypotheses that are plausible candidates for truth cf. (Salis, 2018). A synthetic analysis of the role of models in science is (Gelfert, 2016).

of abduction and chance discovery, which implements a (semantic) dementia care system under the concept of affordance, is proposed.

Perceptual affordances are mainly provided by spatial mappings referred for example to iconic signs such as in the case of pictorial objects for ladies' and men's restrooms: affordances in these cases are provided by cultural conventions. The example of traffic lights is mandatory, red means "stop," yellow means "prepare to stop", and green means "go". Knowledge is needed to recognize these affordances: a triangle does not provide specific geometrical knowledge for people who have no knowledge about Euclidean geometry. In this case, internal knowledge is needed in recognizing suitable affordances.

Abe (2012) introduces the concept of affordance to a dementia care system. Appropriate affordances might offer a certain support to dementia persons understanding/diagnosing (meanings of) objects. In therapy houses, there should be objects and tools that cannot be appropriately used by dementia people. Obviously, affordance changes depending on according on the user's ability. For some individuals, a tree stump will function as a chair, but for other people it will not. Abe observes that

> For semantic dementia persons, it is observed that they did not achieve better performance on a subset of affordable objects when the use of these was compared with a familiarity-matched subset of objects lacking such affordances. Therefore, when we design an environment for dementia persons, it is necessary to consider such unhappy situations. It is necessary to prepare specialized affordances for dementia persons. Even if they can detect affordance, they might not understand what it will emerge (Abe, 2012, p. 166).

This means that we have to improve and so *cure* the eco-cognitive situatedness of dementia persons to favor diagnosticability of the functions of the objects that compose their environments. Something similar is occurring in the case of human and computer interaction: a user-friendly interface might favor users' "discovery processes" of otherwise hidden properties and relationships between entities and data, as already illustrated in the previous sections. Other trivial examples coming from everyday life can be seen in the area of business, in which the era of internet interaction between a customer (user) and goods surely counts: the strategy of information delivery in business is consequently extremely important and curating it is necessary.

In this conceptual framework devoted to deepening the concept of affordance Abe (2010) proposed the concept of curation for chance discovery, providing the case of job in (art) museums and galleries, where curators "remain up-to-date in the scholarly developments within their field(s), conduct original research and develop new scholarship that contributes to the advancement of the body of knowledge within their field(s) and within the museum profession as a whole" (Abe, 2012, p. 168). Curation increases the opportunities of discovering further chances, both explicit (but not previously seen) and implicit, also taking into account the user's interests and situations.

In the case of scientific cognition, curation can also be regarded as an action related to the improvement of interactive "communication" between scientists and their environment (artifacts, available knowledge, appropriate ethical and epistemic rules, collaborators, etc.) In this case, an important aspect of "curation" regards the reaching of appropriate levels of "communication" to favor the best affordances.

In the framework of abduction the following formalization can render the problem of curation in the case of affordances: the most important relationship between an object and meaning is represented by the following formula:

$$Object \cup Object' \cup M \cup affordance \models meaning$$

($M$ is a mapping function from *Object* to *Object'*): "That is, to understand the same meaning of the subset of or similar afforded objects, an additional mapping function $M$ is required. Thus if $M$ can be determined and the usage of *Object* is known, *Object'* can also be understood" (Abe, 2012, p. 169).

Normal people can very simply understand $M$ but this is not the case of dementia persons, who cannot understand and determine $M$ anymore. That is the proper affordance given in the environment is jeopardized and, consequently, the communication between the object and the person missed. Curation restores affordances for dementia persons.

In the case of business we have quoted above the important is to cure the real or digital environment to show rare or hidden but important matters to consumers that in general tend to believe the (filtered) information, that is intentionality, is fundamental on the part of the sellers. In general, curation can be instead not necessarily related to finalized intentions but, more simply, as the job to provide users opportunities to discover chances that should not be explicitly displayed to users, but simply rendered more reachable, according to the user's interests and situations: this consists in providing suitable affordances offered to the agents at play, from the case of customers to the more complicates case of scientists.

Abe provides the clear case of the AED (Automated External Defibrillator) apparatus: if we cannot find an AED apparatus and cannot find or understand the instruction, we cannot save an ill person. Thus two appropriate affordances have to be provided, they should be explicitly displayed and the instruction should be easy to understand. Abe further says that an AED apparatus can be regarded as a "shikake" that functions as a trigger.[18]

Hence, in the case of scientific cognition, a chance is *a rare and novel event* that can trigger discovery:[19] a better and curated environment can be regarded as

[18] (Abe, 2013). According to Matsumura (2012), a shikake is a trigger to start an action or to change a person's mind and behavior to the aim of solving a problem. It sometimes is not the person's will. Thus, as shikakeology teaches, such operation should be conducted implicitly. Matsumura also suggests that shikake should be appropriately designed: the relationship between a problem to solve and a trigger to action should be appropriately designed. Matsumura himself uses the term "affordance" to explain such trigger.

[19] Standard triggers of abductive cognition in science are anomalies (empirical but also theoretical), events or problems not completely predicted by existing theories or events (or problems) partially explained by more than one theory (Aliseda, 2006).

equipped with a shikake. The epistemic environment will have

1. to remain current in the scholarly developments within the scientific field(s) at stake;
2. to provide accessible, transparent, and updated knowledge;
3. to offer the more recent and appropriate technological artifacts, when required;
4. to assume responsibility for the overall care and development of the artifacts, data, structures, laws, levels of commmodification of scientific results, their secreteness and intellectual property, which do not have to impede the knowledge transfer, which characterizes scientific cognition starting from the time of Galileo Galilei;
5. to be characterized by institutional policies and procedures for the care of the reasoning activities of the humans involved and of their collectives, based on accepted professional standards and best practices.

In the following chapters of this book I will illustrate the various dangers and problems that can jeopardize that eco-cognitive openness I think mandatory to favor those discoveries in science that are characterized by that "uberty" Charles Sanders Peirce considered a fundamental quality of the best creative hypotheses.[20]

## 2.4 Protecting the Inferential View of Abduction, Affordances, and Cognitive Niches

A final note has to be added concerning some aspects of the framework of studies about cognition that take advantage of the so-called free energy principle (FEP) (organisms, that are "porous" to the world rather than distanced from it must be acting to minimize long-term average prediction error).[21] In the previous chapter I introduced the distinctions between creative and selective abduction and also the concept of abduction to the best explanation, distinctions which in this chapter have been further complicated by the reference to manipulative abduction, affordances, and cognitive niches, as concepts able to favor the study of the problem of discoverability. FEP is usually illustrated as active in an anti-inferentialist theory, which emphasizes the role in human and animal cognition of the predictive processing (PP), which operates with approximate Bayesian inference. The organism's internal model constantly updates the weighting of the sensory input it collects in the light of the learned predictions of its precisions. Furthermore, dynamic precision weighting guides at the same time both perception and action.

The theory provides a unifying framework in which the difference between selective and creative abduction, abduction to the best explanation, and other kinds of abduction are often described as "deceptive" effects of that "inferentialism"

[20] On uberty cf. this book, Sect. 1.2 of the previous chapter.

[21] Cf. for example the relevant (Clark, 2016).

typical of the "intellectualist, epistemological vernacular" (Hohwy, 2019, p. 198) of philosophical and cognitive science theories: simply, in this framework, inference to the best explanation approximated Bayes. Indeed, to make an example, it is stated that

> Clark applies pressure in the opposite direction: internal processing "accommodates" rather than "explains" the sensory barrage, and since all biological organisms, including single-celled ones, are supposed to engage in prediction-error minimisation, intellectualist, epistemological labels are incongruous. By removing PP from the epistemological, inferential framework, Clark is in a better position to say that PP is servicing utility for the organism rather than aiming at truth. This helps situate PP better in the EEE [embodied, extended, enacted—author note] scheme. (Hohwy, 2019, p. 198)

I think it is possible to nullify the distinctions above concerning the various types of abduction, but this "execution" is valid only in the light of the restricted interpretation of FEP: indeed, this does not mean that the above distinctions become a "deception". EEE theories are not just "rhetoric" at all, as stated by Friston (2019, p. 183). In this last case we face a hostile simplification which is implied by a kind of anti-philosophical and anti-cognitive FEP-based "scientism" (sometimes typical of various reductionist forms of Bayesianism, I have to say). Seeing Peircean abduction in the light of PP and "active inference" is certainly "natural", but we have to remember that in the perspective of that theory we cannot see anymore various aspects of that fresh knowledge about human cognition that instead epistemology and cognitive science profusely supply. If "inferentialism sits poorly with the idea that PP also applies to simpler creatures than humans" (Hohwy, 2019, p. 200) it is also because the concept of inference is considered in a restricted way: a simple referral to Peirce would have provided a richer concept of inference. Indeed, I have already repeatedly said that from Peirce's philosophical perspective, all thinking is in signs, and signs can be icons, indices or symbols: moreover, all inference is a form of sign activity, where the word sign includes "feeling, image, conception, and other representation" (Peirce, 1866–1913, 5.283), and, in Kantian words, all synthetic forms of cognition.[22] After all, I agree with Hohwy (2019, p. 201 and p. 203) that says:

> So it is not out of the question to impute the epistemological and inferential vernacular to simpler creatures [...]. What makes the creatures "simpler" is not that they do not engage in prediction-error minimisation but that their model is simpler. [...] PP is so compelling precisely because it is an account that works only by "closing the loop" between body, mind, and world. The states of the world affect the senses, on the basis of which we perceive the world, but likewise, the states of the brain and the body in turn affect the world, which again causes new sensory input for us. We are causal creatures embedded in the causal structure of the world. This is why PP seems a good marriage for EEE.

In the previous sections I have variously illustrated the role of affordances and abduction in the light of the so-called cognitive niches, aiming at considering

[22] An objection to the idea that it is right to say the PP process is not itself literally inference has been advanced by (Kiefer, 2017). On modeling intuitive processes as inferences, cf. (Hermkes & Mach, 2019). I will further illustrate the Peircean concept of inference in the following chapter.

the conceptual building blocks of discoverability. Still taking advantage of the predictive processing theory (but now in the perspective of extended mind theory), a recent interesting analysis of cognitive niche construction as a way adopted by organisms to create and maintain cause-effect models of their niche as guides for fitness affecting behavior is illustrated in (Constant et al., 2020). We have seen in the previous sections that cognitive functions such as action, perception, and learning take advantage of emergent cognitive niches and their affordances: in the light of the predictive processing theory and of active inference they have to be re-conceptualized as various forms of optimization of organisms' predictive (i.e., generative) models of the world, that obviously derive from construction and optimization of cognitive extensions in a process that is called "extended active inference".

The predictive processing theory is surely interesting and foundational of many aspects of human and non-human animal cognition: also, it can be said that "[...] evolution is just nature's way of performing Bayesian model selection in accord with the free energy principle" (Friston, 2019, p. 183). The problem is that recent research has postulated a "sense of purposefulness" in evolution (Magnani, 2021a), a result that predictive processing theory does not contemplate. Given the fact human *semiotic* brains are engaged in a continuous process of delegation and distribution of cognitive functions to the environment to lessen cognitive limitations, also and especially taking advantage of what I have called "manipulative abduction" and of the construction of cognitive niches, we face a second major participant, after natural selection, in evolution. In sum the role of what has been called "extragenetic information" plays an important role in evolution, potentially altering selective pressures: the process of selection actually selects for "purposeful" organisms that consequently affect evolution.

To defend the inferential view of abduction, affordances, and cognitive niches against the reductionist attitudes of predictive processing theory based on a Bayesian brain we have finally to object that the anti-inferential attitude presents a philosophical cost. Indeed the theory is not interested on *how* organisms generate, explain, and select "things": the semiotic exchange with the environment we have just illustrated, that is related to abduction, to the construction of cognitive niches, and to the role of extra-genetic information, is a continuous generation of *meanings*[23] that deal with purposeful choices of interactions with the environment, even when we are engaged in promoting "discoverability", that are extremely important.

[23] On this issue cf. also (Fabry, 2018), who stresses that research on enculturation and recent work on predictive processing are complementary because predictive processing operates at a sub-personal level and on a physiological time scale of explanation only.

## Summary

In this chapter I have contended that discoverability is of course related to the availability of the appropriate affordances. As we have seen in the previous chapters, humans continuously delegate and distribute cognitive functions to the environment to lessen their limits. They create models, representations, and other various mediating structures, that are thought to be an aid for thinking and sometimes to favor discoverability. In dealing with the exploitation of cognitive resources embedded in the environment I have further stressed the attention to the concept of *affordance*, also to the aim of enriching the recent framework of the so-called EEEE-cognition (extended, embodied, embedded, enacted). In this perspective I have further described human cognition and its evolutionary dimension in terms of environmental situatedness, in which bodies and external—also artifactual—objects and devices play important roles: in this perspective it has been natural to exploit the role of affordances. I have also indicated the function of artifactual cognitive objects and devices in their capacity to extend, modify, or substitute "natural" affordances actively providing humans and many animals with new opportunities for action. Consequently, in my terminology, we have seen that new chances for discovery are not simply information, but they are also "affordances", as environmental anchors that allow us to optimally take advantage of external cognitive resources. Abduction has been still at play, in so far as it illustrates those hypothetical inferences that are operated through actions made up of smart manipulations of the environment to both detect new affordances and to create artefactual external objects that offer new affordances/cues. Finally, I spent some paragraphs to vindicate the importance of the epistemological and cognitive analysis of abduction, affordances, and cognitive niche construction, and their importance for the analysis of discoverability, against some reductionist attitudes generated by certain rigid interpretation of the predictive processing theory based on the idea of "active inference".

## References

Abe, A. (2009). Cognitive chance discovery. In S. C. (Ed.), *Universal access in HCI, part I, HCII2009 (LNCS5614)* (pp. 315–323). Berlin/New York: Springer.

Abe, A. (2010). Curation in chance discovery. In *Proc. ICDM 2010 5th International Workshop on Chance Discovery* (pp. 793–799).

Abe, A. (2012). Cognitive chance discovery: From abduction to affordance. In L. Magnani & P. Li (Eds.), *Philosophy and cognitive science* (pp. 155–172). Cham, Switzerland: Springer.

Abe, A. (2013). Relationship between curation, chance and shikake. *Procedia Computer Science, 22*, 1219–1228. 17th International Conference in Knowledge Based and Intelligent Information and Engineering Systems—KES2013.

Adelman, L., Yeo, C., & Miller, S. (2006). Understanding the effects of computer displays and time pressure on the performance of distributed teams. In A. Kirlik (Ed.), *Human-technology interaction. Methods and models for cognitive engineering and human-computer interaction* (pp. 43–54). Oxford/New York: Oxford University Press.

Adolphs, R. (2006). How do we know the minds of others? Domain specificity, simulation, and enactive social cognition. *Brain Research, 1079*, 25–35.

Aliseda, A. (2006). *Abductive reasoning. Logical investigations into discovery and explanation*. Heidelberg/Berlin: Springer.

Barrett, H. C., & Kurzban, R. (2006). Modularity in cognition: Framing the debate. *Psychological Review, 113(3)*, 628–647.

Bauters, M. (2007). Mediation seen through the sensory eye: An alternative to the "old" and "new" media paradigm. *International Journal of Applied Semiotics, 6(1)* 79–101.

Brunswik, E. (1943). Organismic achievement and envornmental probability. *Psychological Review, 50*, 255–272.

Brunswik, E. (1955). Representative design and probabilistic theory in a functional psychology. *Psychological Review, 62*, 193–217.

Byrne, D., Kirlik, A., & Fick, C. S. (2006). Kilograms matter: Rational analysis, ecological rationality, and closed-loop modeling of interactive cognition and behavior. In A. Kirlik (Ed.), *Human-technology interaction. Methods and models for cognitive engineering and human-computer interaction* (pp. 267–286). Oxford/New York: Oxford University Press.

Clark, A. (2008). *Supersizing the mind. Embodiment, action, and cognitive extension*. Oxford/New York: Oxford University Press.

Clark, A. (2016). *Surfing uncertainty: Prediction, action, and the embodied mind*. New York: Oxford University Press.

Constant, A., Clark, A., Kirchhoff, M., & Friston, K. J. (2020). Extended active inference: Constructing predictive cognition beyond skulls. *Mind and Language*. https://doi.org/10.1111/mila.12330

Degani, A., Shafto, M., & Kirlik, A. (2006). What makes vicarious functioning work? Exploring the geometry of human-technology interaction. In A. Kirlik (Ed.), *Human-technology interaction. Methods and models for cognitive engineering and human-computer inter action* (pp. 179–196). Oxford/New York: Oxford University Press.

Donald, M. (2001). *A mind so rare. The evolution of human consciousness*. London: Norton.

Fabry, R. (2018). Betwixt and between: The enculturated predictive processing approach to cognition. *Synthese, 195*(6), 2483–2518.

Flach, J. M., & Warren, R. (1995). Active psychophysics: The relation between mind and what matters. In J. M. Flach, J. Hancock, P. Caird, & K. Vincente (Eds.), *Global perspectives on the ecology of human-machine systems* (pp. 189–209). Hillsdale, NJ: Erlbaum.

Friston, K. J. (2019). Beyond the desert landscape. In M. Colombo, E. Irvine, & M. Stapleton (Eds.), *Andy Clark and his critics* (pp. 174–190). New York: Oxford University Press.

Gaver, W. W. (1991). Technology affordances. In *CHI'91 Conference Proceedings* (pp. 79–84).

Gelfert, A. (2016). *How to do science with models. A philosophical primer*. Springer.

Gibson, J. J. (1979). *The ecological approach to visual perception*. Boston, MA: Houghton Mifflin.

Hermkes, R., & Mach, H. (2019). An inferential view on human intuition and expertise. In A. Nepomuceno, L. Magnani, F. Salguero, C. Barés-Gómez, & M. Fontaine (Eds.), *Model-based reasoning in science and technology. Inferential models for logic, language, cognition and computation* (pp. 174–276). Cham, Switzerland: Springer.

Hohwy, J. (2019). Beyond the desert landscape. In M. Colombo, E. Irvine, & M. Stapleton (Eds.), *Andy Clark and his critics* (pp. 191–205). New York: Oxford University Press.

Kiefer, A. (2017). Literal perceptual inference. In T. Metzinger & W. Wiese (Eds.), *Philosophy and predictive processing*. Frankfurt: MIND Group.

Kirlik, A. (1998). The ecological expert: Acting to create information to guide action. In *Human Interaction with Complex Systems, (HICS'98) Proceedings* (pp. 22–25.). Fourth Annual Symposium, Dayton, OH. IEEE Computer Society Press.

Kirlik, A. (2006a). Abstracting situated action: Implications for cognitive modeling and interface design. In A. Kirlik (Ed.), *Human-technology interaction. Methods and models for cognitive engineering and human-computer interaction* (pp. 212–226). Oxford/New York: Oxford University Press.

Kirlik, A. (Ed.). (2006b). *Human-technology interaction. Methods and models for cognitive engineering and human-computer interaction*. Oxford/New York: Oxford University Press.

Kirlik, A. (2007). Reiventing intelligence for an invented world. In R. J. Sternberg & D. D. Preiss (Eds.), *Intelligence and technology: The impact of tools on the nature and development of human abilities* (pp. 105–134). Mawhah, NJ: Lawrence Erlbaum Associates.

Kirsh, D. (1995). The intelligent use of space. *Artificial Intelligence, 73*, 31–68.

Kirsh, D., & Maglio, P. (1994). On distinguishing epistemic from pragmatic action. *Cognitive Science, 18*, 513–549.

Laland, K. N., & Brown, G. R. (2006). Niche construction, human behavior, and the adaptive-lag hypothesis. *Evolutionary Anthropology, 15*, 95–104.

Laland, K. N., & Sterelny, K. (2006). Perspective: Seven reasons (not) to neglect niche construction. *Evolution. International Journal of Organic Evolution, 60*(9), 4757–4779.

MacLeod, M., & Nersessian, N. J. (2016). Interdisciplinary problem-solving: Emerging modes in integrative systems biology. *European Journal for Philosophy of Science, 6*(3), 401–418.

Magnani, L. (2006). Mimetic minds. Meaning formation through epistemic mediators and external representations. In A. Loula, R. Gudwin, & J. Queiroz (Eds.), *Artificial cognition systems* (pp. 327–357). Hershey, PA: Idea Group Publishers.

Magnani, L. (2007). *Morality in a technological world. Knowledge as duty*. Cambridge: Cambridge University Press.

Magnani, L. (2009). *Abductive cognition. The epistemological and eco-cognitive dimensions of hypothetical reasoning*. Heidelberg/Berlin: Springer.

Magnani, L. (2021a). Cognitive niche construction and extragenetic information: A sense of purposefulness in evolution. *Journal for General Philosophy of Science, 52*, 263–276.

Magnani, L. (2021b). Computational domestication of ignorant entities. Unconventional cognitive embodiments. *Synthese, 198*, 7503–7532. Special Issue on "Knowing the Unknown" (guest editors L. Magnani and S. Arfini).

Magnani, L., & Bardone, E. (2005). Designing human interfaces. The role of abduction. In L. Magnani & R. Dossena (Eds.), *Computing, philosophy and cognition* (pp. 131–146). London: College Publications.

Magnani, L., & Bertolotti, T. (Eds.). (2017). *Handbook of model-based science*. Switzerland: Springer.

Matsumura, N. (2012). Shikakeology as the dual problem. In *Proc. of the 26th Annual Conference of the Japanese Society for Artificial Intelligence* (pp. 793–799).

McGrenere, J., & Ho, W. (2000). Affordances: Clarifying and evolving a concept. In *Proceedings of Graphics Interface* (pp. 179–186). May 15–17, 2000, Montreal, Quebec, Canada.

Norman, D. A. (1988). *The psychology of everyday things*. New York: Basic Books.

Odling-Smee, F. J., Laland, K. N., & Feldman, M. W. (2003). *Niche construction. The neglected process in evolution*. Princeton, NJ: Princeton University Press.

Peirce, C. S. (1866–1913). *Collected papers of Charles Sanders Peirce*. Vols. 1–6, C. Hartshorne & P. Weiss (Eds.), vols. 7–8, A. W. Burks (Ed.). (1931–1958). Cambridge, MA: Harvard University Press.

Powers, W. T. (1973). *Behavior: The control of perception*. Chicago, IL: Aladine.

Salis, F. (2018). Scientific discovery through fictionally modelling reality. *Topoi*, 1–11.

Scarantino, A. (2003). Affordances explained. *Philosophy of Science, 70*, 949–961.

Stuart, M. T., & Nersessian, N. J. (2018). Peeking inside the black box: A new kind of scientific visualization. *Minds and Machines, 29*(1), 87–107.

Thom, R. (1988). *Esquisse d'une sémiophysique*. Translated by V. Meyer, *Semio Physics: A Sketch*, Addison Wesley, Redwood City, CA, 1990. Paris: InterEditions.

Turing, A. M. (1992). In D. C. Ince (Ed.), *Collected works of Alan Turing. Mechanical intelligence*. Amsterdam: Elsevier.

Turner, P. (2005). Affordance as context. *Interacting with Computers, 17*, 787–800.

Vandi, C. (2007). Beyond metaphors and icons: Towards a perception-action model for graphic user interfaces. Presented at the "Ninth World Congress of IASS/AIS, Communication: Understanding/Misunderstanding", Helsinki, June 2007.

Vicente, K. J. (2003). Beyond the lens model and direct perception: Toward a broader ecological psychology. *Ecological Psychology, 15(3)*, 241–267.

Warren, W. H. (1995). Constructing an econiche. In J. Flach, P. Hancock, J. Caird, & K. J. Vicente (Eds.), *Global perspective on the ecology of human-machine systems* (pp. 210–237). Hillsdale, NJ: Lawrence Erlbaum Associates.

Wells, A. J. (2002). Gibson's affordances and Turing's theory of computation. *Ecological Psychology, 14(3)*, 141–180.

Wheeler, M. (2004). Is language an ultimate artifact? *Language Sciences, 26*, 693–715.

Windsor, W. L. (2004). An ecological approach to semiotics. *Journal for the Theory of Social Behavior, 34(2)*, 179–198.

# Chapter 3
# Eco-Cognitive Openness and Eco-Cognitive Closure

## Locking or Unlocking Strategies? "Knowledge in Motion" Defended

In this chapter, with the help of the concepts of locked and unlocked strategies, abduction, and optimization of eco-cognitive openness, I will describe some central aspects of the cognitive character of reasoning strategies and related heuristics, to the aim of shedding new light on the cognitive aspects of deep learning machines and on how they affect discoverability, and so creativity. Taking advantage of my studies on abduction, I will contend that what I call *eco-cognitive openness* is undermined in the case of famous computational programs such as AlphaGo/AlphaZero, because they are based on *locked abductive strategies*. Unlocked abductive strategies, which are in tune with what eco-cognitive openness requires, qualify those high-level kinds of abductive creative reasoning that are typical of human-based cognition, locked abductive reasoning strategies are instead much simpler than the unlocked ones to be rendered at the computational level. The second part of the chapter emphasizes, taking advantage of the logical and cognitive studies introduced in the previous chapters, the importance of *knowledge in motion*—in multidisciplinary, interdisciplinary, and transdisciplinary scientific research—and its role in favoring high-level kinds of discoverability, and so of abductive results and of the so-called "optimization of eco-cognitive openness and situatedness". The concept of "knowledge in motion" is also the necessary conceptual premise of the analysis of various kinds of "*epistemic irresponsibility*", which are treated in the following chapter.

## 3.1 Locked and Unlocked Strategies in Natural and Artificial Cognition

Go is a game played by human agents and AlphaGo is a computational deep learning program that can play the role of an automatic agent/player, so that competition with humans can become partially computationally determined. Go is

 
L. Magnani, *Discoverability*, Logic, Argumentation & Reasoning 26,
https://doi.org/10.1007/978-3-030-93329-6_3

already an "artificial" game, as it is invented by human beings and, consequently, takes advantage of abstract rules and artifacts, such as the board and other material objects. AlphaGo is artificial too, but a more complicated fruit of the technological creativity of few human beings.[1] However, we have to remember that also "natural cognitive games", so to speak, can be contemplated. For example, as I have already illustrated in the previous chapter when delineating manipulative abduction, a strategic human cognition not only refers to propositional aspects concerning acts performed through written and spoken language, but it is also active in a distributed cognition environment, in a kind of "game" in which models, artifacts, internal and external representations, sensations, and manipulations play a central function: imagine the pre-linguistic cognitive "natural game" between humans and their surroundings, in which "unlocked" strategies (see below) are at play, such as the phenomenological tradition has illustrated, exactly involving embodied and distributed systems, and visual, kinesthetic, and motor sensations (Magnani, 2018).

What counts here is that in the natural games the strategies are *unlocked* because, even if local constraints are always at play in the interaction humans/environments, no predetermined background is established. On the contrary, what happens in the case of human-made "artificial games" such as Go, or in the case of their computational counterpart, such as AlphaGo/AlphaZero? In these two last cases, the involved cognitive strategies are *locked*, and are related to a situation of "eco-cognitive closure", as I will describe in the following paragraphs. As the reader can guess, different kinds of diagnosticability and discoverability are related to locked and unlocked strategies.

Let us abandon the problem of the prelinguistic cognitive abductive strategies which are at play in a natural interaction—natural game—between humans and their prepredicative surroundings (for example, splendidly studied by Husserl, 1931) and let us concentrate on the cognitive abductive strategies that are at play in the artifactual case of the moves that are occurring in the adversarial game Go with two players and their respective changing surroundings, which in this case are basically formed by board, stones, and possible artifactual assisting accessories. In this game, analogously to the case of the natural processes, we still obviously find the role of visual, kinesthetic, and motor sensations and actions (such as in the natural games between humans and their prepredicative surroundings I have just quoted), but also the strong function of visual, iconic, and propositional artificial representations, anchored to the human-made "meanings" (both internal and external) which gave birth to the game and which characterize its features and rules.

[1] AlphaGo Zero is a version of DeepMind's Go software AlphaGo; the recent AlphaZero further enriches AlphaGo Zero, cf. https://en.wikipedia.org/wiki/AlphaZero.

### 3.1.1 Reading Ahead

A fundamental strategy we immediately detect in artificial games such as Go, which is necessary for proficient and smart tactical play, is the capacity to *read ahead*, as the Go players usually say. Reading ahead is a practice of generating groups of anticipations that aim at being very robust and complex (either serious-minded or intuitive) and that demand the consideration of

1. clusters of moves to be adopted and their potential outcomes. Of course, the moves embedded in these clusters do not have to be "discovered" because they just have to be "selected" among a finite set of possible moves already available (obviously, moves do not have to be invented/discovered): in this sense we can say we have to *diagnose* the ones that can be a useful and efficacious part of a suitable cluster. The cluster of moves instead either can be selected in an encyclopedia of clusters of moves already available or to be discovered/invented in the sense that it can be excogitated for the first time, as original and never seen. The available scenario at time $t_1$, exhibited by the board, represents an adumbration[2] of a subsequent potential more profitable scenario at time $t_2$, which indeed is abductively credibly hypothesized: in turn, on the basis of these excogitated scenarios, one more "selective" abduction is actuated, which—consistently and believably—activates a particular move that can lead to an envisaged more fruitful scenario.
2. Possible countermoves to each move.
3. Further chances after each of those countermoves. It seems that some of the smarter players of the game can read up to 40 moves ahead even in hugely complex positions.

In a book published in Japan, related to the description of various strategies that can be exploited in Go games, Davies emphasizes the role of "reading ahead":

> The problems in this book are almost all reading problems. [...] they are going to ask you to work out sequences of moves that capture, cut, link up, make good shape, or accomplish some other clear tactical objective. A good player tries to read out such tactical problems in his head before he puts the stones on the board. He looks before he leaps. Frequently he does not leap at all; many of the sequences his reading uncovers are stored away for future reference, and in the end never carried out. This is especially true in a professional game, where the two hundred or so moves played are only the visible part of an iceberg of implied threats and possibilities, most of which stays submerged. You may try to approach the game at that level, or you may, like most of us, think your way from one move to the next as you play along, but in either case it is your reading ability more than anything else that determines your rank (Davies, 1995, p. 6).

Further abductive strategies that are usefully adopted by human players in the game Go are for instance related to "global influence, interaction between distant

[2] The word belongs to the Husserlian philosophical lexicon (Husserl, 1931) I have analyzed in its relationship with abduction in (Magnani, 2009, chapter four).

stones, keeping the whole board in mind during local fights, and other issues that involve the overall game. It is therefore possible to allow a tactical loss when it confers a strategic advantage".[3]

The material and external configurations of the various scenarios (which are composed by sensible objects—stones and board) that characterize artificial games are the fruit of a cognition "sedimented"[4] in their embodiment, after the starting point of their creation and subsequent uses and modifications. The cognitive tools that are related to the application of both the game allowed rules and the individual inferential talents owned by the two players, strategies, tactics, heuristics, etc. are sedimented in those material objects (artifacts, in this case) that become *cognitive mediators*:[5] for example they orient players' inferences, transfer information, and provoke reasoning chances. Once represented internally, the external subsequent scenarios become object of mental manipulation and new ones are further made, to the aim of producing the next most successful move.

It is important to note again that these strategies, when actuated, are certainty characterized by an extended variety, but all are "locked", because the elements of each scenario are always the same (what changes is merely the number of seeable stones and their dispositions in the board), in a finite and stable framework (no new rules, no new objects, no new boards, etc.) These strategies are devoid of the following feature: they are not able to recur to reservoirs of information *different* from the ones available in the fixed given scenario. It is important to add a central remark: of course, the "human" player can enrich and fecundate his strategies by referring to internal resources not necessarily directly related to the previous experience with Go, but with other preexistent skills belonging to disparate areas of cognition. This is the reason why we can say that the strategies of a "human" player present a less degree of closure with respect to the automatic player AlphaGo. In humans, strategies are locked with respect to the external rigid scenario, but more open, and so unlocked, with respect to the mental field of reference to previous wide strategic experiences or to hypotheses about the behavior of the adversary: however, everything concurs to the management of an unalterable scenario. In this case, the human unlocked cognitive system may still sadly lack sufficient computational resources to win against the computational tool,

[3] Cf. Wikipedia, entry Go (game) https://en.wikipedia.org/wiki/Go_(game).

[4] An expressive adjective still used by Husserl (1978).

[5] This expression, I have extendedly used in (Magnani, 2001), is derived from Hutchins, who introduced the expression "mediating structure", which regards external tools and props that can be constructed to cognitively enhance the activity of navigating. Written texts are trivial examples of a cognitive "mediating structure" with clear cognitive purposes, so mathematical symbols, simulations, and diagrams, which often become "epistemic mediators", because related to the production of scientific results. "Language, cultural knowledge, mental models, arithmetic procedures, and rules of logic are all mediating structures too. So are traffic lights, supermarkets layouts, and the contexts we arrange for one another's behavior. Mediating structures can be embodied in artifacts, in ideas, in systems of social interactions [...]" (Hutchins, 1995, pp. 290–291) that function as an enormous new source of information and knowledge.

notwithstanding the fact in computational programs such as AphaGo and deep learning systems, the strategic reservoir cannot—at least currently—take advantage of that mental openness and flexibility typical of human beings. The repertoire is merely formed/learned to play the game by checking data of thousands of games, and no other sources. Of course, as I have already anticipated, it is this gap that affects discoverability.

I have also to say that the notion of cognitive locked strategy I am referring to here is not present in and it is unrelated to the usual technical categorizations of game theory. Fundamentally, in combinatorial game theory, Go can be technically illustrated as zero-sum (player choices do not increment resources available-colloquially), perfect-information, partisan, deterministic strategy game, belonging to the same class as chess, checkers (draughts) and Reversi (Othello). Moreover, Go is bounded (every game has to end with a victor within a finite number of moves and time), strategies are obviously associative (that is in function of board position), the format is of course non-cooperative (no teams are allowed), positions are extensible (that is they can be represented by board position trees).[6]

## 3.2 Locked Abductive Strategies Defeat the Eco-Cognitive Openness

In my research I have recently emphasized (Magnani, 2017, chapter seven) the *knowledge enhancing* character of abduction, which I have also sketched in the first chapter of this book. This means that in this case we are dealing with abductive reasoning strategies that grant successful and highly creative outcomes. The knowledge enhancing feature regards several kinds of newly generated knowledge of various novelty degrees, from that new knowledge about a suffering patient we have abductively accomplished in medical diagnosis (a case of selective abduction, as no new biomedical knowledge is created, just new knowledge about a person) to the new knowledge developed in scientific discovery, which many epistemologists celebrated, for example, Paul Feyerabend in *Against Method*, as I have already emphasized in chapter one of this book (Feyerabend, 1975). In the case of an artificial game such as Go, the knowledge activated thanks to an intelligent choice of already available strategies or thanks to the invention of novel strategies and/or heuristics must also be considered a result of knowledge enhancing abduction.

I strongly contend that, to arrive to "uberous" selective or creative optimal abductive results, in the case of scientific discovery, useful strategies must be applied, but it is also needed to be in presence of a cognitive environment marked by what I have called *optimization of eco-cognitive situatedness*, in which eco-cognitive openness is fundamental (Magnani, 2016). This feature of the cognitive environment is especially needed in the case of strong creative abduction, that is

[6] Cf. Wikipedia entry Go (game) https://en.wikipedia.org/wiki/Go_(game).

when the kind of novelty is not restricted to the case of the "simple" successful diagnosis. In Sect. 3.3 I will illustrate in more detail that to favor good creative and selective abduction reasoning and so diagnosticability and discoverability, strategies must not be "locked" in an external restricted eco-cognitive environment, such as in a scenario characterized by fixed defining rules and finite material aspects, which would function as cognitive mediators able to constrain agents' reasoning.

At this point, it is valuable to supply a brief sketch of the concept of *eco-cognitive openness*, I have already explained in chapter one: I emphasized the importance in good abductive cognition to hypotheses of what has been called *optimization of situatedness*. Let us recall what is the meaning of the expression optimization of situatedness thanks to the example of scientific reasoning. We already remarked that abductive cognition is very important in scientific reasoning because it refers to that activity of creative hypothesis generation which characterizes one of the more important aspects of rational knowledge. To get abductive results in science, the "situatedness" of the related cognitive activities is strongly connected with eco-cognitive aspects, related to the contexts in which knowledge is "traveling": in the case of scientific abductive cognition (but also in other abductive cases, such as diagnosis) to favor the solution of an inferential process the situatedness also has to be characterized by the richness of the flux of information, which has to be maximized. This maximization aims at that optimization of situatedness, which, as I quoted above, can only be made possible by a *maximization of changeability* of the basic data which inform the abductive cognitive process: inputs have to be maximally enriched, rebuilt, or modified and the same has to occur with respect to the knowledge applied during the hypothetical reasoning process. Obviously, the aim is to have at disposal a favorable "cognitive environment" in which available data can become *optimally positioned*.[7]

In summary, we said that abductive processes to hypotheses—in a considerable quantity of cases, for example in science—are highly *information-sensitive*, and face with a flow of information and data uninterrupted and appropriately promoted and enhanced when needed (of course also thanks to artifacts of various kinds). This means that also from the psychological perspectives of the individuals the epistemological openness in which knowledge channeling has to be maximized is fundamental.

## 3.3 Locking Strategies Restricts Creativity

Optimization of situatedness is related to unlocked strategies and grants a good degree of both diagnosticability and discoverability. Locked strategies, such as the ones active in Go game, AlphaGo/AlphaZero, and other computational AI systems and deep learning devices, do not favor the optimization of situatedness. Indeed,

[7] I have furnished more cognitive and technical details to explain this result in (Magnani, 2016).

I have already contended above that, to obtain certain good kinds of creative and selective abductions, reasoning strategies must not be "locked" in bounded eco-cognitive surroundings (that is, in scenarios designed by fixed defining rules and finite material objects which would play the role of the so-called cognitive mediators). In this perspective, a poor scenario is certainly responsible for the minimization or, more simply, reduction, of the eco-cognitive openness and it is the structural consequence of the constitutive organization of the game Go (and also of Chess and other games), as I have already described in Sect. 3.1.1. I have said that in the game Go stones, board, and rules are rigid and so totally predetermined; what instead is undetermined are the strategies and connected heuristics that are adopted to defeat the adversary in their whole process of application.[8]

As I have already said, the available strategies and the adversary's ones are always *locked* in the fixed scenario: you cannot, during a Go game, play for a few minutes Chess or adopt another rule or another unrelated cognitive process, affirming that that weird part of the game is still appropriate to the game you agreed to play. You cannot decide to change the environment at will so *unlocking* your strategic reasoning, for example because you think this will be an optimal way to defeat the adversary. Furthermore, your adversary cannot activate at his discretion a process of eco-cognitive transformation of that artificial game. On the contrary, in the example of scientific discovery, the scientist (or the community of scientists) frequently recur to disparate external models and change their reasoning strategies[9] to produce new analogies or to favor other cognitive useful procedures (prediction, simplification, confirmation, etc.) to enhance the abductive creative process.

The case of scenarios in human scientific discovery precisely represents the counterpart of the ones that are poor from the perspective of their eco-cognitive openness. Indeed, in these last cases, the reasoning strategies that can be endorsed (and also created for the first time), even if multiple and potentially infinite, are *locked* in a determined perspective where the components do not change (the stones can just diminish and put aside, the board does not change, etc.) I would say that in scenarios in which strategies are locked, in the sense I have explained, an *autoimmunization* (Magnani & Bertolotti, 2011; Arfini & Magnani, 2015) is active, that constitutes the limitations that preclude the application of strategies that are not related to "pre-packaged" scenarios, strategies that would be foreigners to the ones that are strictly intertwined with the components of the given scenario. Remember I already said that these components play the role of *cognitive mediators*, which anchor and constrain the whole cognitive process of the game.

[8] Of course, many of the strategies of a good player are already mentally present thanks to the experience of several previous games.

[9] Many interesting examples can be found in the various parts of the recent *Handbook of Model-Based Science* (Magnani & Bertolotti, 2017).

To summarize and further explain (by linking the problem of locked and unlocked strategies to the various cases of selective and creative abduction):

1. Contrarily to the case of high level "human" creative abductive inferences such as the ones expressed by scientific discovery or other examples of special exceptional intellectual results, the status of artificial games (and of their computational counterpart) is very poor from the point of view of the non-strategic knowledge involved. We are dealing with stones, a modest number of rules, and one board. When the game progresses, the shape of the scenario is spectacularly modified but no unexpected cognitive mediators (objects) are appearing: for example, no diversely colored stones, or a strange hexagonal board. On the contrary, to continue with the example of high levels creative abductions in scientific discovery (for example, in empirical science), first of all the evidence is extremely rich and endowed with often unexpected novel features (not only due to modifications of aspects of the "same things", as in the case of artificial games). Secondly, the flux of knowledge at play is multifarious and is related to new analogies, thought experiments, models, imageries, mathematical structures, etc. that are rooted in heterogeneous disciplines and fields of intellectual research. In sum, in this exemplary case, we are facing with a real tendency to a status of optimal eco-cognitive situatedness (further details on this kind of creative abduction are furnished in Magnani, 2015, 2016, 2017).
2. What happens when we are dealing with selective abduction (for example in medical diagnosis)? How locked and unlocked strategies affect diagnosticability? First of all, evidence freely and richly arrives from several empirical sources in terms of body symptoms and data mediated by sophisticated artifacts (which also change and improve thanks to new technological inventions). Second, the encyclopedia of biomedical hypotheses in which selective abduction can work is instead locked,[10] but the reference to possible new knowledge (locally created or externally available) is not inhibited, so the diagnostic inferences can be enhanced thanks to scientific advancements at a first sight not considered. Third, novel inferential strategies and linked heuristics can be created and old ones used in new surprising ways but, what is important, strategies are not locked in a fixed scenario. In sum, the creativity that is occurring in the case of human selective abduction is poorer than the one active in scientific discovery, but richer than the one related to the activity of the locked reasoning strategies of the Go game and AlphaGo, I have considered above.
3. In Go (and similar games) and in deep learning systems such as AlphaGo, in which strategies and heuristics are "locked", these are exactly the only part of the game that can be improved and rendered more fertile: strategies and related heuristics can be used in a novel way and new ones can be invented, even if always locked. Anticipations as abductions (which incarnate the activities of "reading ahead") just affect the modifications and re-grouping of the same

[10] It is necessary to select from pre-stored diagnostic hypotheses.

elements. No other types of knowledge will increase; all the rest remains stable.[11] Of course, this dominance of the strategies is the quintessence of Go, Chess, and other games, and also reflects the spectacularity of the more expert moves of the human champions. However, it has to be said that this dominance is also the reason that explains the fact the creativity at stake is even more modest than the one involved in the higher cases of selective abduction (diagnosis). I will soon illustrate that this fact is also the reason that explains why the smart strategies of Go or Chess games can be more easily simulated, for example with respect to the inferences at play in human scientific discovery, by recent artificial intelligence programs, such as the ones based on deep learning.[12]

The reader does not have to misunderstand me: I do not mean to minimize the relevance of creative heuristics as they work in Go and other board games. John Holland already clearly illustrated (1995, 1997) that board games such as checkers, as well as Go, are wonderful cases of "emerging" cognitive processes, where potentially infinite strategies favor exceptional games: even if simply thanks to a few rules regulating the moves of the pieces, games cannot be predicted starting from the initial configurations. While other cases of emerging cognitive processes (I have indicated the example of human scientific discovery) characterize what can be called "vertical" creativity (that is, related to unlocked strategies), board games are examples of "horizontal" creativity: even if board games are circumscribed by locked strategies that constrain the game, "horizontal" creativity can show astonishing levels of creativity and skillfulness. We already said that these extraordinary humans skills have been notably appropriated by artificial intelligence software (see below the last paragraphs of this section): the example given in this chapter is the one of AI deep learning heuristics that were able to *learn from* human games, such in the case of AlphaGo, and also from games autonomously played, such in the case of AlphaZero. What are the remaining most important effects which derive from these computational AI programs equipped to concretize cognitive abductive inferences characterized by "locked" strategies?

At the beginning of this chapter, I have illustrated some amazing performances of Google DeepMind's program AlphaGo against human players and the fact the system showed to be able to "create" unconventional moves—never played by humans—thus building new strategies, a fact that undoubtedly favors the attribution to the system of actual "human" capacities, and we have to add, better than the ones of the more skilled humans. AlphaGo instructed itself to play "attending" (so to speak) thousands of games played by human beings thanks to "reinforcement learning", which refers to that activity of self-playing of the machine (the machine plays against itself) to further feed and adapt its own neural networks.

---

[11] Obviously, for example, new rules and new boards can be proposed, so realizing new types of game, but this chance does not jeopardize my argumentation.

[12] On the results of the so-called—non-human—automated scientific discovery in AI cf. (Magnani, 2009, chapter two), section 2.7 "Automatic Abductive Scientists".

In the passage I am reporting below, Colheo and Thompsen Primo seem to corroborate my contention regarding the fact I consider these kinds of human cognitive performances considerably simple to be reproduced at the computational level with AI programs as AlphaGo because they are dealing with what I have called in this chapter *locked reasoning strategies*. In summary, a kind of background reason that explains this easiness would be that this type of human cognition is less creative than others, even if it is so striking when optimally realized by very smart human intelligent subjects.

> Let us compare the key ideas behind Deep Blue (Chess) and AlphaGo (Go). The first program used values to assess potential moves, a function that incorporated lots of detailed chess knowledge to evaluate any given board position and immense computing power (brute force) to calculate lots of possible positions, selecting the move that would drive the best possible final possible position. Such ideas were not suitable for Go. A good program may capture elements of human intuition to evaluate board positions with good shape, an idea able to attain far-reaching consequences. After essays with Monte Carlo tree search algorithms, the bright idea was to find patterns in a high quantity of games (150,000) with deep learning based upon neural networks. The program kept making adjustments to the parameters in the model, trying to find a way to do tiny improvements in its play. And, this shift was a way out to create a policy network through billions of settings, i.e., a valuation system that captures intuition about the value of different board position. Such search-and-optimization idea was cleverer about how search is done, but the replication of intuitive pattern recognition was a big deal. The program learned to recognize good patterns of play leading to higher scores, and when that happened it reinforces the creative behavior (it acquired an ability to recognize images with similar style) (Coelho & Thompsen Primo, 2017).

I think humans with their biological brains do not have to feel mortified by these extraordinary skillful capacities of the AI programs. Unfortunately, given the present worldwide status of mass media, other magnificent human performances in various fields of creativity, much more creative than the ones related to locked strategic reasoning, are unable to reach the global echo AlphaGo/AlphaZero gained. Indeed, human-more-skillful-abductive creative capacities, related to unlocked strategies, as I have tried to demonstrate in this chapter—still cognitively beautiful—are not sponsored by Google, which is a herculean corporation that can easily obtain the attention of not only the monocultural media of our age but also of the social networks: many human beings are more easily impressionable by the "miracles" of AI, robotics, and information technologies, than by prodigious knowledge results of human beings-like-us, too often out of sight (after all—ça va sans dire—also AI traditional programs and AI deep learning systems have been created by humans. . . )

Google managers also think that AI deep learning programs similar to AlphaGo could be exploited to help science resolve important real-world problems in health-care but also in other fields. This would be a good research program. Google seems to also expect to implement some business thanks to commercialization of new deep learning AI powers to collect appropriate information and generate abductions in some advantageous fields. Simply checking the Wikipedia entry DeepMind (https://

en.wikipedia.org/wiki/DeepMind),[13] [DeepMind is a British artificial intelligence company instituted in September 2010 and took by Google in 2014, the company created the AlphaGo program], the following non-contested passage is reported, concerning the so-called "NHS data-sharing controversy":

> In April 2016, New Scientist obtained a copy of a data-sharing agreement between DeepMind and the Royal Free London NHS Foundation Trust. The latter operates three London hospitals where an estimated 1.6 million patients are treated annually. The agreement shows DeepMind Health had access to admissions, discharge, and transfer data, accident and emergency, pathology and radiology, and critical care at these hospitals. This included personal details such as whether patients had been diagnosed with HIV, suffered from depression, or had ever undergone an abortion in order to conduct research to seek better outcomes in various health conditions. A complaint was filed to the Information Commissioner's Office (ICO), arguing that the data should be pseudonymized and encrypted. In May 2016, New Scientist published a further article claiming that the project had failed to secure approval from the Confidentiality Advisory Group of the Medicines and Healthcare products Regulatory Agency. In May 2017, Sky News published a leaked letter from the National Data Guardian, Dame Fiona Caldicott, revealing that in her "considered opinion" the data-sharing agreement between DeepMind and the Royal Free took place on an "inappropriate legal basis". The Information Commissioner's Office ruled in July 2017 that the Royal Free hospital failed to comply with the Data Protection Act when it handed over personal data of 1.6 million patients to DeepMind.

Even if based on what I called in this chapter locked strategies, and thus far from the highest levels of human creativity, AI deep learning systems and various other programs can also offer chances for business and good integration in the market. I think epistemologists and logicians have to monitor the use of these AI devices (of course when less transparent than the natural and limpid—and so stupefying—performance of AlphaGo in games against humans). Recent research in the field of epistemology, cognitive science, and philosophy of technology[14] illustrate that good AI software, which surely furnishes a big new chance for opportunity and data analytics, can be transmuted in a tool that does not respect epistemological and/or ethical rigor. For example, in the case regarding the computational exploitation of big data, outcomes can inadvertently lead to epistemologically unacceptable computer-discovered correlations (instead possibly good from a commercial perspective), but these tools are sometimes—unfortunately—seriously illustrated as aiming at replacing tout-court human-based scientific research as a guide to understanding, prediction, and action. Calude and Longo say: "Consequently, there will be no need to give scientific meaning to phenomena, by proposing, say, causal relations, since regularities in very large databases are enough: 'with enough data, the numbers speak for themselves' " (Calude & Longo, 2017, p. 595). Unfortunately, some "correlations appear only due to the size, not the nature, of data. In 'randomly' generated, large enough databases too much information tends to behave like very little information" (Calude & Longo, 2017, p. 595). I agree with these authors: we cannot treat some spurious correlations as results of deep scientific creative

[13] Date of access 8 of April, 2021.

[14] Relatively recent bibliographic references can be found in my book (Magnani, 2007).

abduction, but just as trivial generalizations, even if reached with the help of sophisticated artifacts.[15] I cannot further deepen the problems regarding issues connected to the impact of computational programs on ethics and society, that will be treated in the following chapters. In the first part of this chapter I limited myself to deal with cognitive, logical, and epistemological aspects to the aim of introducing the distinction between locked and unlocked strategies and its meaning with respect to intelligent computation.

## 3.4 Governing Diagnosticability and Discoverability: Confinement, Filter Bubbles, and Ignorance

The recent studies concerning the so-called "cognitive confinement" also deal with aspects that can be referred to the dichotomy between locked and unlocked strategies. Cognitive confinement (Werner, 2021) refers "to a contingent, yet relatively stable state of being structurally or systematically unable to gain information from an environment, determined by patterns of interaction between the subject and the world" and it is considered as a pathologized form of cognitive niche.[16] It is obvious that confinement affects human and nonhuman animal abductive cognition. In chapter two we have described the role of affordances, that emerge as the outcome of specific patterns of subject/world interaction: "[. . . ] cognitive confinement is a state of being systematically blocked from affordances or even disafforded, as it were, and so provided with disfigured affordances due to the way in which the subject's environment (cognitive niche) is set up by the subject-world interactions" (Werner, 2021). A situation of embodied and ecological ignorance, as the author also calls it.

In the perspective of this book, we can further see confinement as the situation in which it is the eco-cognitive situatedness that is circumscribed and limited. The complexity of the interactions between the agent and the environment governs the kind of cognitive confinement that is occurring. The example of the so-called "filter bubbles" can be of help: access to online information—in the so-called virtual cognitive niches—is constrained by algorithms that filter content in accordance with the subject's previously detected preferences or biases, and this of course diminishes the likelihood of getting new or unexpected information. In my view, filter bubbles are special cases of a more general phenomenon of cognitive confinement—as

[15] On this problem and other negative epistemological use of computational programs, cf. the recent (Calude & Longo, 2017) and the last chapter of this book.

[16] The concept of cognitive niche is explained in the previous chapter, Sect. 2.1.1.

Werner says—related to processes of "ignorance-making". We can face cases in which

1. ignorance/confinement is—explicitly or implicitly—but intentionally built: for example, filter bubbles in social networks create collectives of people who are ignorant because they are almost totally isolated from facts and opinions that would for example change their political orientation. Of course this is a case related to the violence and dangers of propaganda with respect to the good health of liberal democracy, which degrades because of the lack of widespread well-informed decisions and deliberations. The paradox is related to the fact that filter bubbles endanger modern democracies exactly in a time in which people access a huge quantity of data than ever before;
2. ignorance/confinement is the unintentional objective effect of some changes that occur in social equilibria pertaining to specific cognitive niches: in the remaining sections of this chapter and in chapters four and five we will analyze the case of recent "epistemic niches" that jeopardize (or tend to jeopardize) discoverability in some areas of science such as biomedicine. In these cases dealing with evidence and data available is not about any amount of data, but rather about the value or utility of the data afforded to someone. Hence, the case is about the variety and quality of the affordances (which constitute a fraction of the potential ones) one can get from one's environment, evaluated in terms of their use in solving specific problems, such as for example in creative cases.

### *3.4.1 Disfigured Affordances, Adversarial Abduction, and Fake Diagnosticability (and Discoverability)*

Following Werner (2021) we can also consider the role of the so-called disfigured or misleading affordances.

> A different situation occurs when the subject does receive affordances which I call disfigured, which means, roughly, the perception of something that seems to stand for a possible course of action, e.g., the actions that someone might associate with bent sticks, whereas in fact the actions in question are not a real option. A different way of putting the latter is to say that the subject is *dis*afforded.

I have to add that also deception as the activity of intentionally misleading other agents' hypothetical inferences has been fruitfully recently analyzed (Rovetta, 2020): deception is, in this case, analyzed as an intentional manipulation of others' hypothetical inferences so that they will accept the false or disadvantageous abduced hypothesis. This topic is also related to the role of the so-called adversarial cognition in intelligence service, in which the main aim is "to detect and exploit adversarial affordances", as environmentally situated opportunities for knowledge and action defined by a mutual relationship between adversaries (Forsythe, 2017b). Of course it is observed that the most salient and characteristic feature of an intelligence system, common to all intelligence phenomena, is abductive cognition. In these cases it is diagnosticability more than discoverability at play, and various ways of jeopardizing selective abduction (diagnosis) of other human beings are active, so as they consider the abductive performances reliable and good, but they are bad results in the third-person perspective of the deceiver that collaborated to (usually intentionally) generate them to get some advantages.[17]

A further related issue concerns the fact that chances as conditions of diagnosticability and discoverability can further be faked. It seems intuitive to think that the suppression of chance-faking contexts is an important activity of chance curation, as I have illustrated in Sect. 2.3 of the previous chapter of this book. (Magnani & Bertolotti, 2013) considered the amazing example of bullshit as a source of fake chances, but at the same time also—rarely—as a fertile ground for learning and developing some creative intuitions: a rich cognitive niche may benefit from certain kind of fake chances—which should therefore not be inhibited—whereas a poorer niche might not benefit from these fake aspects, and therefore the preclusion of fake chances is an important act of chance curation (Magnani & Bertolotti, 2013).[18]

## 3.5 Maximizing Discoverability in Scientific Discovery: The Centrality of Eco-Cognitive Openness

In the first chapter of this book we have illustrated that in science the optimization of eco-cognitive openness and situatedness also concerns the maximization of the information that has to be made available, which epistemologically characterizes the tradition of modern science. This fact favors a new attention to the current fragile status of human abductive cognition in the scientific enterprise. Is the current situation of scientific research facing problems that jeopardize the optimization of situatedness needed by good abductive cognition? I plan to provide an illustration of some negative aspects that have to be stressed when dealing with the current challenges to the creative productiveness of human abductive cognition in science. Indeed, I will stress the hot problem of the contemporary emergence of disparate kinds of "epistemic irresponsibility", related to the commodification and commercialization of science, the marketing of technoscientific products, the impoverishment of the so-called "epistemic niches", which show that human fruitful

[17] The problem of evading detection in the information age, related to the problem of surveillance, crypsis, and resistance is analyzed in (Forsythe, 2017a): Surveillance and countersurveillance are seen as competing strategies for detecting—that is abducing—and concealing particular kinds of affordance.

[18] On scientific misconceptions, as beliefs that are considered to be false in light of accepted scientific knowledge, seen as generated inside a system of inferences/explanations that do not match a current scientific theory (as not merely isolated false beliefs) cf. (Dellantonio & Pastore, 2021).

abductive cognition in science is increasingly assaulted and jeopardized, and at the same time human creativity possibly endangered.

In this book I am especially dealing with the important case of science, in which the optimization of eco-cognitive situatedness has to respect various constraints. In the last two chapters of this book I will stress that constraints, methods, cognitive virtues, etc., which characterize scientific cognition certainly substantiate limitations of the related cognitive behaviors but at the same time they are clearly finalized to the maintenance of the optimization/maximization of eco-cognitive openness, that solely can grant the flourishing of abductive scientific creativity.

### 3.5.1 "Knowledge in Motion" Defended: Favoring Scientific Abduction through the Maximization of Eco-Cognitive Openness

In Sect. 1.5 of the first chapter I have illustrated the role in abductive cognition of what I have called the *optimization of situatedness*. Of course situatedness is related to eco-cognitive aspects: to favor the solution of the abductive problem input and output of the formula (3) given at p. 16 have to be thought as *optimally positioned*. Not only, this optimality is also made possible by a *maximization of changeability* of both input and output; again, not only inputs have to be enriched with the possible solution but, to do that, other inputs have usually to be changed and/or modified. This changeability first of all refers to a wide psychological/epistemological openness.

To take another example of how this openness can lead researchers to high results, we can think of the adaptability to change required to make use of a serendipitous event in scientific research. The notion of serendipity has been used in the past fifty years to describe the unintentional, accidental, and lucky discoveries made both in lay environments and in the scientific framework. In the latter category some game-changing events for the history of science stand out as unquestionable instantiations of serendipity: Fleming's discovery of penicillin (cited by Slowiczek & Peters, 2007; Kakko & Inkinen, 2009; McCay-Peet & Toms, 2015), Antoine Henri Becquerel's finding of X-rays (cited by Cannon, 1940; Allen, Erdelez, & Marinov, 2013; Rosenman, 1988), Watson and Crick's reflection on the alpha-helix structure of DNA (cited by Darbellay, Moody, Sedooka, & Steffen, 2014; Campanario, 1996; Copeland, 2017), and the list grows longer as the historiographical narrations (and the contemporary scientific reports) become more detailed.

Indeed, the fortune of the term serendipity is connected to the recent acknowledgement of the role of chance in scientific processes (Cannon, 1940; Austin, 2003; Kipnis, 2005; Catellin, 2014) as well as the importance of a skillful perspective that scientists need to enact in order to exploit those lucky events (Allen et al., 2013; Copeland, 2017). For this reason, not only serendipity is a topic of growing fame in the panorama of sociology and philosophy of science, but it can well be used to discuss the maximization of changeability required to improve discoverability in scientific endeavors. Indeed, what characterizes serendipity is a game-changing property that allows the agents involved to make highly creative inferences, such as *trans-paradigmatic abductions* (Arfini, Bertolotti, & Magnani, 2020).

Trans-paradigmatic abduction (Hendricks & Faye, 1999)—already introduced above in chapter one, Sect. 1.3—is basically a concept that defines the possibility to leap from one array of explanatory ideas to another by examining and discussing a compelling hypothesis that belongs to the latter. Thus, we could use the concept of trans-paradigmatic abduction to explain the necessary openness to new perspectives that scientific research requires. To make a concrete example, we can consider the case of Barry Marshall and Robin Warren, who proposed the theory that peptic ulcers are caused by bacterial infections after having discovered a new kind of bacteria in the stomachs of people with gastritis. As Thagard (1998, p. 114) states, "The discovery of *Helicobacter pylori*, and the formation of the hypothesis that it routinely colonizes human stomachs, are the result of a combination of serendipity, questioning, and search." This hypothesis was initially rejected by the medical community because the stomach was previously believed to be a sterile environment where bacteria could not survive long enough to damage it.

Thus the generation of the hypothesis (which emerged from a lucky observation that led to a series of them) did not just provide a new explanation for the phenomenon of peptic ulcers, but a new viewpoint on the biological composition and balance of the stomach. This new point of view has been initially proposed when the inference was generated to make a reason for an observation that could not be explained within the previous theoretical framework. At that moment the trans-paradigmatic abduction operated the game-changing switch that post hoc has been recognized by the medical community when referring to that serendipitous process.

As many authors in the literature of abduction affirm, the mere consideration of the new hypothesis does not initiate the process of abduction: the agent needs to cognitively and epistemically invest in the hypothesis, actually embracing it, in order to claim that the inferential process has been conducted (Gabbay & Woods, 2005). Thus, the conjunction of the lucky observation, the elaboration of the trans-paradigmatic abduction, and its testing compose the game-changing serendipitous process, which obviously can be enacted only by embracing a maximization of changeability in the scientific environment. So, at this point we could ask: is the current situation of scientific research facing problems that jeopardize the optimization of situatedness needed by good abductive cognition? I plan to illustrate in the following sections some negative aspects that have to be stressed when dealing with the current challenges to the creative productiveness of abductive cognition in science.

It is in this perspective that we will clearly see that in many fields of current scientific and technological research we are not simply dealing with ethical problems, such as for example it is the case of research on information technology, engineering, biomedical sciences, etc., but also with "epistemological" problems.

Some processes related to what I call "epistemic irresponsibility" are currently active: this fact presents a new task for philosophy of science, now more directly social relevant with respect to the whole community of citizens. Philosophy of science, traditionally related to the analysis of scientific knowledge, methods, and cognitive procedures, now can also address the problem of trust in the relations between science and society (Whyte & Crease, 2010) and at the same time the crucial obligation of detecting the situations in which epistemic rigor is jeopardized in a way that renders scientific rationality annihilated or at best—so to speak—"faked".[19]

In what I have called epistemic niches,[20] that is cognitive niches that are expressly built to host and favor scientific rationality and scientific reasoning, mindful dissent is free and promoted, but also respected and taken into account, secrecy abolished, exchange of ideas encouraged and always active, reliable evidence respected, publicly recognized standards for the evaluation of research established, etc. It is in this perspective that we can clearly see science as a valued-laden enterprise, a perspective in which there is not a strict distinction between the moral and the epistemic: practicing science requires a moral commitment to follow certain adopted methods, rules, cognitive behaviors. In the absence of these aspects an epistemic niche is simply no more epistemological, that is devoted to the progress of science.

Longino contends that to make possible scientific research we need an appropriate social dimension, that is communities that approximate an ideal free communication of ideas, such as that the one illustrated by John Stuart Mill. Biddle makes a clear criticism to this perspective (Biddle, 2009): Longino presupposes a conception of the individual that undermines her own claim that scientific knowledge is necessarily social. She would think that certain social arrangements are just helpful ways of producing scientific knowledge, but not necessary. I simply think wide individual freedom from any kind of constraint, which is granted in certain communities (for example in the epistemic niches) is extremely important. Of course this character of the community needs special care to be maintained and preserved: for example in the case of scientific creative abduction, in which that

[19] A recent analysis concerning the fact that commercial interests can have a negative impact on the trustworthiness of science is given in (Fernández Pinto, 2020). The author contends that commercial interests explicitly impact research in ways that render results unreliable, for example, when patent cases of scientific fraud or misconduct occur policy makers and the general public are well-justified to distrust such results. Hence, commercialized science is also considered the central factor in the erosion of trust in science: commercial interests can promote the production of consensus or dissent that is industry-friendly, but not epistemically acceptable, and commercial interests generate a narrow understanding of trustworthiness that leads the industry to claim that their research is reliable when it is not.

[20] See above chapter two, this book, Sect. 2.1.1.

situation of eco-cognitive openness (related to the need for the optimization of eco-cognitive situatedness) I described in (Magnani, 2017) is crucial.[21]

This does not mean that other—contextual—values play no role in science: Popper emphasized that metaphysical or theological beliefs orient the kinds of theoretical frameworks available to scientists and moral, economical, and political values orient the choice of one theoretical framework over another. The essential point, however, is that given that contextual values do play an inevitable role in scientific reasoning, the ideal of value neutrality should be rejected. Actually, we can say that certainly science is oriented by values but these values do not contaminate its epistemic purity and its *neutrality*: just to make an example, we have a lot of scientific research about biological cell and less research about animal cognition, and this is definitely due to some external constraints that oriented research in a certain direction, but in both cases epistemic purity is intact. Researchers in both fields need an optimization of eco-cognitive openness, which can be reached thanks to both the individual attitude and the social conditions of maximal elimination of obstacles to free communication of ideas and criticism.[22]

Biddle continues his critique to Longino saying that her emphasis on what he calls the "unencumbered selves" produces a false idea of individuals who have complete freedom to choose their own aims, values, and conceptions, and this would not reflect the attitude of real scientists (Biddle, 2009). I think that he is wrong, at least in the perspective of the creative aspects of science. Creative abductive scientists are individuals who have to actually be open to everything and who live in a situation of "openness to all perspectives: no claim or belief can be held immune to criticism" (Longino, 2002, p. 159). This does not mean they are not socially located and that this freedom has to be seen active in other aspects of scientific research, unrelated to creative abduction: in these last cases individuals need not be—so to speak—completely open-minded, because they belong to precise epistemic niches which provide established rules and heuristics, and also methodologies and evaluative criteria, making them capable of questioning and evaluating certain beliefs (and not all of their beliefs).

In sum, high levels of eco-cognitive openness regard the creative aspects of scientific enterprise. In this perspective we do not have to see any tension between Longino's employment of Mill's political philosophy and her account of the social character of scientific enterprise. Individual scientists can be considered open to everything at the cognitive level of abductive creativity, but at the same time they live in social epistemic niches characterized by established paradigms, research programs, exemplars, etc., which provide firm rules and criteria of various kinds able to govern a great part of scientific cognitive processes, as already richly illustrated, during the postpositivistic renaissance of philosophy of science, by Kuhn, Lakatos,

---

[21] In the first chapter I have also emphasized that in creative abduction irrelevance and implausibility have to be exculpated (see also Magnani, 2017, chapter six): to favor free production and communication of ideas, without barriers, they remain fundamental to foster innovation.

[22] Cf. also the following chapter, Sect. 4.1.1.

Feyerabend (Kuhn, 1962; Lakatos, 1970; Feyerabend, 1975) and by their many followers. In conclusion, individual scientists are certainly to be considered as "advocates for particular approaches, paradigms, or research programs" as Biddle says (2009, p. 622)—Lakatos famously wrote of "the rationality of a certain amount of dogmatism" (1970, p. 175)—but *also* individuals who are open to everything when they face with the problem of abducing new ideas.

Hence, sustaining discoverability, that is a situation of optimization of eco-cognitive openness, is not only related to the individual attitude of the good "creative" scientist but also to the specific character of the social, political, and economical environment in which a certain scientific research is embedded, that must constitute a good background, appropriate to the end.[23]

Hence, let us come back to the problem of the suitable conditions of eco-cognitive openness necessary to creative abduction. Even if many contemporary scientific research problems transcend disciplinary boundaries imposed by institutions of higher education, narrow-minded specialization certainly represents a limitation to the flourishing of abductive cognition: this aspect represents a first challenge to the exercise of good abductive cognition. A contrary example if furnished by HIV/AIDS studies, I will illustrate below. This contemporary prosper creative scientific research has been strongly characterized by the cross-fertilization of ideas in a kind of situation of "knowledge in motion".

Let us take a look at some recent aspects of scientific research related to health problems, which demonstrate the importance for abduction of the optimization of eco-cognitive openness. Recent studies in HIV/AIDS do not completely fit into the research fields typically found in the excess of blind disciplinarity. Researchers and funding agencies have instead devoted increasing attention to projects that span multiple disciplines. Taking advantage of scientometric computational techniques (the so-called "topic models" of over 9000 abstracts from two prominent journals on HIV/AIDS, from their inception through the end of 2010) Light and Adams, in an article eloquently entitled "Knowledge in motion: the evolution of HIV/AIDS research" (see Adams & Light, 2014; Light & Adams, 2016 and) describe how research programs in this area moved through various states of integration including disciplinarity, multidisciplinarity, interdisciplinarity and transdisciplinarity. In my perspective these aspects are all related to the actual promotion of optimization of that eco-cognitive openness which characterizes creative abductions in science.

Multidisciplinary research engages any single research area from the vantage point of multiple disciplines, this approach poses little challenge to the specialization which characterizes the more disciplinary-based approaches. Light and Adams

---

[23] Social epistemologists always emphasized the role in scientific rationality and objectivity of the institutions and processes of knowledge acquisition and production (Longino, 1990; Kitcher, 1993). Recently Reiss—manly referring to biomedical research—noted the importance of not restricting normative judgments about how to organize research to ethical aspects alone (for example, discriminating against diseases of ethnic minorities and the poor in the Western world, or the diseases of the global poor) but also to the epistemic components of research when for example epistemic decency is disrespected (Reiss, 2010).

(2016, pp. 1229–1230) report that, following the suggestions given by the U.S. National Academy of Sciences "[...] interdisciplinary research is a more integrative form: interdisciplinary research consists of research conducted by teams or individuals that integrates information, data, techniques, tools, perspectives, concepts, and/or theories from two or more disciplines or bodies of specialized knowledge to advance fundamental understanding or to solve problems whose solutions are beyond the scope of a single discipline or field of research practice". Finally, "transdisciplinarity is characterized by relatively complete elimination of the salience of disciplinary boundaries in the scientific enterprise. Within the literature conceptually clarifying the typology of many-disciplined science, transdisciplinarity's boundaryless science arises via two quite different processes. The first form comes from a reorganization of scientific efforts—starting with the identification of questions, and extending straight through the strategies employed in answering them—around problem-based areas. That is, research is no longer bound up within disciplinary confines".

Fortunately, research on HIV/AIDS of the last decades has been characterized by a balance between specialization and cross-pollination of ideas. The integration across disciplinary boundaries is fundamentally dynamic:

> HIV/AIDS research appears to have grown more multidisciplinary also consistent with increased specialization, after an initial period of more interdisciplinary integration. In other words, after an initial period of high cross-fertilization and joint problem-solving, the researchers returned to their silos working on overlapping topics in a more disciplinary fashion. As evidenced by the case of HIV/AIDS research, integrated programs are dynamic and situate themselves into different states of organization (Light & Adams, 2016, p. 1245).

### 3.5.2 *Marketing Technoscientific Results*

The example of HIV/AIDS studies I have just illustrated exemplifies a case of what we can call *epistemic responsibility*: scientists and the various agents involved in decision making concerning their eco-cognitive environments all concur in building policies, attitudes, intellectual habits, social and economical institutions, appropriate artifacts, able to promote responsible epistemic actions: what happens when "knowledge in motion" is not favored? To have an idea of the relatively recent situation of the whole scientific research in U.S. it is sufficient to quote the following passage:

> We are currently witnessing profound changes in the way in which scientific research in the United States is organized. In 1964, 30.8% of U.S. R&D was funded by industry, while 66.8% was funded by the federal government. The years between 1964 and the present have witnessed an almost exact reversal; in 2004, 63.8% of national R&D was funded by industry, while only 29.9% was funded by the federal government. [...] Growing numbers of both university and government scientists are developing financial relationships with private corporations, and many university scientists are starting their own companies, simultaneously playing the roles of academic researchers and entrepreneurs. The result of these changes is that commercial considerations are exerting increasing influence within

the practice of science, in both universities and governmental laboratories (Biddle, 2011, pp. 245–246).

(Lazonick, 2007, p. 11) draws a similar picture but looking at the data about internal R&D (i.e. R&D made in house by pharma companies). He makes the example of Pfizer reporting that

> In response to complaints that U.S. drug prices are at least twice those in any other country, Pfizer and other U.S. pharmaceutical companies have argued that the profits from these high prices—enabled by a generous intellectual-property regime and lax price regulation—permit more R&D to be done in the United States than elsewhere. Yet from 2003 through 2012, Pfizer funneled an amount equal to 71% of its profits into buybacks, and an amount equal to 75% of its profits into dividends. In other words, it spent more on buybacks and dividends than it earned and tapped its capital reserves to help fund them. The reality is, Americans pay high drug prices so that major pharmaceutical companies can boost their stock prices and pad executive pay.

In few words, scientists are now forced into areas that have the potential to be profitable, first of all they are much less free than in the past to decide for themselves which questions to study. The introduction of market values—particularly encouraged by the neoliberal politics—for example into the U.S. practice of science aims at resulting in continued scientific progress (and so discovery) but also in a huge economic and social growth. It is well-known that this policy also favors deregulation, by removing the government-imposed barriers, the flow of information between private and public sectors, the sharing of expertise and the transfer of scientific research into marketable products.

Unfortunately, allowing market forces to guide some fundamental areas of research, such as the biopharmaceutical one, has generated a degradation of the epistemic quality: "a sacrifice of epistemic standards at the altar of profit" (Biddle, 2011, pp. 246). Knowledge is no more "in motion". The U.S. Congressional Patent and Trademark Amendments Act of 1980—known as the Bayh-Dole Act—allowed universities and private corporations to patent the results of publicly-funded research, while the Stevenson-Wydler Technology Innovation Act of 1980 (later on revised in 1986) allowed for the patenting of results obtained in government laboratories (Biddle, 2011). Prior to these acts, results deriving from privately-funded research could be privately appropriated, whereas inventions resulting from publicly-funded research obviously remained in the public domain. We are facing with a clear "[...] proprietary treatment of research results, with the commercial interest in secrecy overriding the public's interest in free, shared knowledge" (Brown, 2000, p. 1701). In sum, the new "obsession" regarding commercialization generally, and patenting in particular, has obviously been generated not by the scientific community, but rather by political and economic decisions (Biddle, 2012).[24]

[24] I think that things are even worse as the neoliberals nowadays push (public) universities to actively pursue the so-called "technology transfer"—which means that universities need to produce knowledge that is quickly transferable to firms so that they can exploit it commercially. So researchers are affected by the fact they are also constantly put at pressure by this "technology

This change has profoundly affected the eco-cognitive environment of important parts of science: in the following chapter I will illustrate many aspects of the current eco-cognitive situations which put in danger discoverability and so the productivity of scientific discovery and consequently jeopardized and are jeopardizing the richness of human abductive cognition in science. I will take advantage of the prototypical case of biomedical and pharmaceutical research, in which the dangers of an objective situation of epistemic irresponsibility are already more than remarkable. I consider this example prototypical because it represents a clear kind of epistemic irresponsibility, which might very likely spread all over many other areas of science and of technological research and which surely is already seriously threatening the survival of areas not directly marketable, such as ample parts of humanities, mathematics, and science, instead fundamental to preserve and "feed" western civilization.

A first example regards the role of positive and negative freedom of research in our era. Knowledge in motion, which is necessary to promote that "optimization/maximization of eco-cognitive situatedness" needed by fostering good human abductive creative reasoning is only promoted by the existence of the so-called positive freedom, as freedom to act in particular ways or to achieve particular results. Unfortunately, neoliberals sustain a short-sighted view of the nature of freedom, as freedom from external obstacles, which is not sufficient to guarantee substantial freedom in scientific research. Commercialized scientists have less obstacles—neoliberals weakened the state institutions that now no longer can satisfactorily rule behaviors and circumstantiate distinctions between public and private interests—but are oppressed by pre-established agendas. Not only, paradoxically, and contrarily to all expectations—a new big obstacle appeared: once "commercialized" and dependent upon the private sectors, the scientists are no more free to communicate research results freely, so that negative freedom, otherwise defended, is instead violated. For example, this outcome affects both sponsored biopharmaceutical industry and agricultural biotechnology, where for-profit corporations have a tendency to only consider research that is beneficial to them financially and to avoid research that is not (Elliott, 2012).[25] A second example relates to the fact that "for-profit corporations have a history of biasing studies to increase the likelihood of obtaining desired results" (Biddle, 2014, p. 17). Big pharma tend to withhold information that might call the safety or efficacy of their drugs into question.

Financial considerations directly influence the choice of research problems, the choice of methods, and the interpretation of data, and the ways in which research is disseminated and rendered public, and indirectly influence research by generating an

transfer" obsession—which often sounds absurd in those countries which invest ridiculous amounts of money in research.

[25] This author also illustrates, on the contrary, how strategies of information dissemination are often implemented for example just to dissipate public suspicions (for example in the case of the genetic modification (GM) of crops.)

atmosphere in which researchers fear to study particular topics. The overall effect resorts to a kind of inhibition of various kinds of research and at the same time of its potential creativity). (Biddle & Leuschner, 2015, p. 274) observe: "A recent study (Shulman, 1999) found that more than one-third of recently published articles produced by University of Massachusetts scientists had one or more authors who stood to make money from the results they were reporting. That is, they were patent holders or had some relationship, for example, as board members, to a company that would exploit the results. The financial interests of these authors were not mentioned in the publications" (Brown, 2000, p. 1701). It is also interesting to quote the so-called "publication bias", due to the fact that some research results are not published, usually the scientific publications which indicate "problems" with respect to some important patented—or to be patented—products.[26]

To the eyes of a researcher in logic, epistemology, and cognitive science like me it seems perverse to see realized the commercialization of many scientific processes and results, for example the fact that few owners can control the "product" of a scientific discovery/invention. It seems perverse because we all know that science is a communal enterprise, which grew generation after generation so that all the new results are in themselves also indebted to the past achievements. Some new technologies would have been impossible without previous so-called "theoretical" discoveries—which of course it was impossible to market—that now seem to mean nothing in the perspective of the current commercialization process. Geniuses and personalities are obviously important but their contribution has always been strongly tied to the human social settings in which they worked and to the rich cultural heritage arrived from the past, that is to something related to the cultural "commons". In this perspective, just to make a simple example, how to justify exclusive property rights, when not a painting or a romance but a GMO is at stake, appears to be extremely problematic.

## Summary

In this chapter, with the help of the concepts of locked and unlocked strategies, abduction, and eco-cognitive openness (and of its optimization), I have described some central aspects of the cognitive character of reasoning strategies and related heuristics, to the aim of shedding new light on the cognitive aspects of deep learning machines and on how they affect discoverability, and so creativity. Taking advantage of my studies on abduction, I have contended that what I call *eco-cognitive openness* is undermined in the case of famous computational programs

[26] We have to say that also in economics/management disciplines, studies that find not-statistically significant relationships (i.e. they find neither positive nor negative statistical relationships) are not even published in journals. This is amazing because studies that found non-statistically significant results actually could potentially be highly interesting, especially for the fact that they might reverse assumptions (van Hilten, 2015).

such as AlphaGo/AlphaZero, because they are based on *locked abductive strategies*. Instead, unlocked abductive strategies, which are in tune with what eco-cognitive openness requires, qualify those high-level kinds of abductive creative reasoning that are typical of human-based cognition, locked abductive reasoning strategies are instead much simpler than unlocked ones to be rendered at the computational level. The second part of the chapter emphasized, taking advantage of the logical and cognitive studies introduced in the previous chapters, the importance of *knowledge in motion*—in multidisciplinary, interdisciplinary, and transdisciplinary scientific research—and its role in favoring in high-level kinds of abductive cognition the so-called "optimization of eco-cognitive openness and situadedness". The concept of "knowledge in motion" is also the necessary conceptual premise of the analysis of various kinds of "*epistemic irresponsibility*", which will be treated in the following chapter.

## References


Adams, J., & Light, R. (2014). Mapping interdisciplinary fields: Efficiencies, gaps and redundancies in HIV/AIDS research. *PLoS One, 19(12)*. https://doi.org/10.1371/journal.pone.0115092

Allen, C. M., Erdelez, S., & Marinov, M. (2013). Looking for opportunistic discovery of information in recent biomedical research: A content analysis. *Proceedings of the American Society of Information Science and Technology, 50*(1), 1–22.

Arfini, S., Bertolotti, T., & Magnani, L. (2020). The antinomies of serendipity. How to cognitively frame serendipity for scientific discoveries. *Topoi, 39*, 939–948.

Arfini, S., & Magnani, L. (2015). An eco-cognitive model of ignorance immunization. In L. Magnani, P. Li, & W. Park (Eds.), *Philosophy and cognitive science II. Western & eastern studies* (Vol. 20, pp. 59–75). Switzerland: Springer.

Austin, J. H. (2003). *Chase, chance, and creativity: The lucky art of novelty*. Cambridge, MA: The MIT Press.

Biddle, J. B. (2009). Advocates or unencumbered selves? On the role of Mill's political liberalism in Longino's contextual empiricism. *Philosophy of Science, 76*(5), 612–623.

Biddle, J. B. (2011). Bringing the marketplace into science: On the neoliberal defense of the commercialization of scientific research. In M. Carrier & A. Nordmann (Eds.), *Science in the context of application* (pp. 245–269). Cham: Springer.

Biddle, J. B. (2012). Tragedy of the anticommons? Intellectual property and the sharing of scientific information. *Philosophy of Science, 79*(5), 821–832.

Biddle, J. B. (2014). Can patents prohibit research? On the social epistemology of patenting and licensing in science. *Studies in History and Philosophy of Science Part A, 45*(1), 14–23.

Biddle, J. B., & Leuschner, A. (2015). Climate skepticism and the manufacture of doubt: Can dissent in science be epistemically detrimental? *European Journal for Philosophy of Science, 5*(3), 261–278.

Brown, J. R. (2000). Privatizing the university–the New Tragedy of the Commons. *Science, 290*(5497), 1701–1702.

Calude, C. S., & Longo, G. (2017). The deluge of spurious correlations in big data. *Foundations of Science, 22*(3), 595–612.

Campanario, J. (1996). Using citation classics to study the incidence of serendipity in scientific discovery. *Scientometrics, 37*(1), 3–24.

Cannon, W. B. (1940). The role of chance in discovery. *The Scientific Monthly, 50*(3), 204–209.

Catellin, S. (2014). *Sérendipité. Du conte au concept*. Paris: Seuil.

Coelho, H., & Thompsen Primo, T. (2017). Exploratory apprenticeship in the digital age with AI tools. *Progress in Artificial Intelligence, 1*(6), 17–25.

Copeland, S. (2017). On serendipity in science: Discovery at the intersection of chance and wisdom. *Synthese* 1–22. Online First.

Darbellay, F., Moody, Z., Sedooka, A., & Steffen, G. (2014). Interdisciplinary research boosted by serendipity. *Creativity Research Journal, 26*(1), 1–10.

Davies, J. (1995). *Tesuji. Elementary go series* (Vol. 3). Tokyo: Kiseido Publishing Company.

Dellantonio, S., & Pastore, L. (2021). Ignorance, misconceptions and critical thinking. *Synthese, 198*, 7473–7501.

Elliott, K. (2012). Selective ignorance and agricultural research. *Science, Technology, and Human Values, 38*(3), 328–350.

Fernández Pinto, M. (2020). Commercial interests and the erosion of trust in science. *Philosophy of Science, 87*(5), 1003–1013.

Feyerabend, P. (1975). *Against method*. London-New York: Verso.

Forsythe, S. (2017a). Evading detection in the information age on surveillance, crypsis and resistance. Technical Report.

Forsythe, S. (2017b). On affordance and the adversary: Intelligence in theory and practice. Under graduate Dissertation, War Studies and Philosophy, Faculty of Social Science and Public Policy Department of War Studies, King's College London.

Gabbay, D. M., & Woods, J. (2005). *The reach of abduction: Insight and trial*, Amsterdam: North Holland: Elsevier.

Hendricks, F. V., & Faye, J. (1999). Abducting explanation. In L. Magnani, N. J. Nersessian, & P. Thagard (Eds.), *Model-based reasoning in scientific discovery* (pp. 271–294). New York: Kluwer Academic/Plenum Publishers.

Holland, J. H. (1995). *Hidden order*. Reading, MA: Addison-Wesley.

Holland, J. H. (1997). *Emergence: From chaos to order*. Oxford: Oxford University Press.

Husserl, E. (1931). *Ideas. General introduction to pure phenomenology* [First book, 1913]. Translated by W. R. Boyce Gibson. London: Northwestern University Press.

Husserl, E. (1970). *The crisis of European sciences and transcendental phenomenology* [1954]. Translated by. D. Carr. London: George Allen & Unwin and Humanities Press.

Husserl, E. (1978). The Origin of Geometry (1939). In J. Derrida (Ed.), Edmund Husserl's "The Origin of Geometry" (pp. 157–180). Translated by D. Carr and originally published in Husserl, 1970, pp. 353–378. Stony Brooks, NY: Nicolas Hays.

Hutchins, E. (1995). *Cognition in the wild*. Cambridge, MA: The MIT Press.

Kakko, I., & Inkinen, S. (2009). Homo creativus: Creativity and serendipity management in third generation science and technology parks. *Science & Public Policy, 36*(7), 537–548.

Kipnis, N. (2005). Chance in science: The discovery of electromagnetism by H.C. Oersted. *Science & Education, 14*(1), 1–28.

Kitcher P. (1993). *The advancement of science: Science without legend, objectivity without illusions* Oxford: Oxford University Press.

Kuhn, T. S. (1962). *The structure of scientific revolutions*. Second expanded edition, 1970. Chicago: University of Chicago Press.

Lakatos, I. (1970). Falsification and the methodology of scientific research programs. In I. Lakatos & A. Musgrave (Eds.), *Criticism and the growth of knowledge* (pp. 365–395). Cambridge, MA: The MIT Press.

Lazonick, W. (2007). The big idea. Profits without prosperity. Stock buybacks manipulate the market and leave most Americans worse off. *Harvard Business Review, 3* 1–11.

Light, R., & Adams, J. (2016). Knowledge in motion: The evolution of HIV/AIDS research. *Scientometrics, 17*, 1227–1248.

Longino, H. E. (1990). *Science as social knowledge: Values and objectivity in scientific inquiry*. Princeton: Princeton University Press.

Longino, H. E. (2002). *The fate of knowledge*. Princeton University Press.

Magnani, L. (2001). *Abduction, reason, and science. Processes of discovery and explanation*. New York: Kluwer Academic/Plenum Publishers.

Magnani, L. (2007). *Morality in a technological world. Knowledge as duty*. Cambridge: Cambridge University Press.

Magnani, L. (2009). *Abductive cognition. The epistemological and eco-cognitive dimensions of hypothetical reasoning*. Heidelberg/Berlin: Springer.

Magnani, L. (2015). The eco-cognitive model of abduction. 'Απαγωγή now: Naturalizing the logic of abduction. *Journal of Applied Logic, 13*, 285–315.

Magnani, L. (2016). The eco-cognitive model of abduction. Irrelevance and implausibility exculpated. *Journal of Applied Logic, 15*, 94–129.

Magnani, L. (2017). *The abductive structure of scientific creativity. An essay on the ecology of cognition*. Cham, Switzerland: Springer.

Magnani, L. (2018). Playing with anticipations as abductions. strategic reasoning in an eco-cognitive perspective. *Journal of Applied Logic—IfColog Journal of Logics and their Applications, 5*(5), 1061–1092. Special Issue on "Logical Foundations of Strategic Reasoning" (guest editors W. Park and J. Woods).

Magnani, L., & Bertolotti, T. (2011). Cognitive bubbles and firewalls: Epistemic immunizations in human reasoning. In L. Carlson, C. Hölscher, & T. Shipley (Eds.), *CogSci 2011, XXXIII Annual Conference of the Cognitive Science Society*. Boston MA: Cognitive Science Society.

Magnani, L., & Bertolotti, T. (2013). Selecting chance curation strategies: Is chance curation related to the richness of a cognitive niche? *International Journal of Knowledge and System Science 4*(1), 50–61.

Magnani, L., & Bertolotti, T. (Eds.). (2017). *Handbook of model-based science*. Switzerland: Springer.

McCay-Peet, L., & Toms, E. (2015). Investigating serendipity: How it unfolds and what may influence it. *Journal of the Association for Information Science and Technology, 66*(7), 1463–1476.

Reiss, J. (2010). In favour of a Millian proposal to reform biomedical research. *Synthese, 177*(3), 427–447.

Rosenman, M. F. (1988). Serendipity and scientific discovery. *The Journal of Creative Behavior, 22*(2), 132–138.

Rovetta, F. F. (2020). Framing deceptive dynamics in terms of abductive cognition. *Pro-Fil, 21*(1), 1–10.

Shulman, S. (Ed.). (1999). *Owning the future. Staking claims on the knowledge Frontier*. New York: Houghton Mifflin.

Slowiczek, F., & Peters, P. M. (2007). Discovery, chance, and the scientific method. Access excellence classic collection. Retrived from http://www.accessexcellence.org/AE/AEC/CC/chance.html

Thagard, P. (1998). Ulcers and bacteria I: Discovery and acceptance. *Studies in History and Philosophy of Science Part C: Studies in History and Philosophy of Biology and Biomedical Science, 29*(1), 107–136.

van Hilten, L. (2015). Why it's time to publish research "failures". *Science Communication*. https://www.elsevier.com/connect/scientists-we-want-your-negative-results-too

Werner, K. (2021). Cognitive confinement: Theoretical considerations on the construction of a cognitive niche, and on how it can go wrong. *Synthese, 198*, 6297–6328.

Whyte, K. P., & Crease, R. (2010). Trust, expertise and the philosophy of science. *Synthese, 177*(3), 411–425.

# Chapter 4
# Jeopardizing Discoverability

## Epistemic Irresponsibility: Human Creative Abduction Attacked

In this chapter I will analyze some important aspects of the organization of Research and Development (R&D) in the case of biopharmaceutical companies, which represent a prototypical situation of what I call impoverished epistemic niches. At least in this case we clearly see a challenge to the epistemic integrity of modern science. Taking advantage of the logical and cognitive studies illustrated in the previous chapters, which emphasize the crucial role played in abductive cognition by the so-called "optimization/maximization of eco-cognitive openness and situadedness", this chapter will describe the hot problem of the current emergence of disparate kinds of "*epistemic irresponsibility*" that strongly jeopardize discoverability.[1] Interesting cases related to the commodification and commercialization of science, marketing of technoscientific products, impoverishment of the so-called epistemic niches are illustrated, which show that human fruitful abductive cognition in science is increasingly assaulted and jeopardized, and at the same time human creativity seriously endangered.

[1] In this chapter I prefer to adopt the expression epistemic irresponsibility (instead of epistemological irresponsibility), because I attribute to the adjective epistemic a restricted meaning. "Epistemic" pertains to scientific knowledge or the conditions for acquiring it, which involve eco-cognitive situations in which not only scientists but also other economical, political, and institutional agents directly or indirectly intertwined with scientific research are involved. Instead "epistemological" also expressly refers to the philosophical community of epistemologists, who in general are not affected by epistemic irresponsibility, but by other problems such as, for example, philosophical irrelevance or scholasticism.

 
L. Magnani, *Discoverability*, Logic, Argumentation & Reasoning 26,
https://doi.org/10.1007/978-3-030-93329-6_4

## 4.1 Jeopardizing Human Creative Abduction Through Impoverished Epistemic Niches

### *4.1.1 Epistemic Irresponsibility I: Expensive Drugs Now and the Undisciplined Commodification of Abduction in Science*

The problem I will address in this subsection concerns other threats to the flourishing of human creative abductive processes when performed in high technological business-oriented environments, in which what has been called a "commercialization of science" (Brown, 2000; Biddle, 2007) is active, consequent to increasing dependence of universities on industry and philanthropy. It is not only important the simple social and political fact that in the background there are forces that are transforming the universities into servants of industry, rather we also need to address the "epistemological" consequences of this processes on creativity and its survival in scientific practice.

In general the effects of private sectors in orienting scientific (and not scientific)[2] research are very well-known. When orienting values are "private" or "philanthropic" values, less neutral than the public ones, the situation is clear. Brown illustrates this aspect, not related to our problem of the threat to human abductive reasoning, but worthy to be reported, taking advantage of an amazing consideration, which also indicates that promoting a certain kind of knowledge is at the same time an act of "selecting ignorance" (Tuana, 2006; Proctor & Schiebinger, 2008; Elliott, 2012b):[3]

> [...] financial institutions donated a very large sum to a Canadian university economics department to study "the effects of high taxation on productivity". The results may influence government policy. In such cases, the public and its political decision-makers get information only of a certain kind, because there is no private, well-funded foundation called The Consortium of Single Mothers on Welfare that bestows similar massive funding to discover the effects of poverty on the development of children. Public policy decisions should be based on a variety of sources of information, but the privatization of research means that one point of view – guess who's? – will tend to prevail. Publicly funded science, though far from perfectly serving all interests, has at least a chance of serving more (Brown, 2000, p. 1701).[4]

---

[2] In this case one of the more patent effect has been the decimation of humanities, because neoliberals continue to ignore that the utility of a liberal arts education resorts to the development of general analytic and writing abilities which are fundamental in industry, government, and administration. It has to be said that also various "scientific" departments, not inclined to do patentable and marketable work, disappeared.

[3] I have already introduced the notion of selective ignorance in Magnani (2017, chapter two).

[4] An extreme paradoxical case of selective ignorance case is illustrated by Elliott (2012b, p. 335): "Another crucial way in which the choice of questions can contribute to selective ignorance is by steering society's responses to a problem toward scientific or technical fixes rather than social or political solutions. Thus, before scientists even begin to formulate their own specific

Further, Brown nicely describes the bad effects of a philanthropic group which, for example, provides funds to a university for studying science and religion, inclining the research to show that science and religion are in perfect harmony (and giving the students this impression), also creating the premises for a potential slippery slop "If religious foundations can fund science and religion courses, then why can racist foundations not fund race and IQ courses?" (Brown, 2000, p. 1701). Another example regards biopharmaceutical research: in approaching a health problem the way that leads to market a new drug will be favored, another, for example involving diet, exercise, or environmental control disregarded. Vaccine research itself has declined also because these drugs are not profitable as the ones for clear chronic conditions, a situation recently totally reversed thanks to the corona virus pandemic. The cognitive atmosphere that in these cases is created is very epistemologically poor: the paucity of rival theories and approaches always deprives science of the fundamental aspect of comparative evaluation, so strongly emphasized by Popper and Lakatos.

It seems that in various areas of research the integrity of scientific knowledge is challenged by heavily funded interest groups who aim at influencing research to serve their needs: "These influences can take a wide variety of forms: funding specific areas of research, designing studies so as to obtain particular results, suppressing undesirable findings and harassing scientists that disseminate them, paying scientists to produce opinionated letters or commentaries or review essays, and developing strategic public-relations campaigns via the mass media [...]. These interest-group strategies can affect scientific language as well as other areas of science" (Elliott, 2012a, p. 16).

However, we have to illustrate a more focused concern: I contend that the commercialization of science is detrimental to both scientific—especially with respect to the *creative* parts of scientific cognition—and social progress. For example the situation created in the case of biomedical research by the so-called "Tragedy of Anticommons" is strikingly anti-innovative. Biddle contends it is particularly interesting this tragedy, proposed by Heller & Eisenberg (1998), for biomedical research: "According to this thesis, the proliferation of intellectual property rights upstream (i.e., over the results of basic research included engineered living organisms that are not 'natural', so to speak) creates a series of obstacles to downstream (i.e., applied) research and product development; the result is that upstream patenting not only fails to incentivize the development of innovative

questions about a research topic, a significant and value-laden decision may already have been made tacitly by citizens or policy makers—namely, the decision to treat the problem primarily as a set of questions for scientists and engineers to address. For example, the IAASTD [International Assessment of Agricultural Knowledge, Science and Technology for Development] emphasizes that there are a number of non-technical solutions to agricultural problems: increasing poor farmers' security of access to land, promoting more transparent and high functioning markets, and providing better social safety nets." On the problem of ignorance with respect to the women's health movement and feminists' commitment to uncovering the ways women's bodies have been ignored, also by science, such as for example in the case of drug companies that decided research in the area of hormonal contraception for men would not be profitable, cf. (Tuana, 2006).

products, it also discourages it" (Biddle, 2012, p. 821). For example, patents in areas such as DNA diagnostics seem to jeopardize the development of certain kinds of knowledge and certain technological capabilities that have the potential to save lives. It is possible that, for many scientists—even unconsciously—projects that have little commercial potential or little prospect of obtaining a patent are never seriously considered. Empirical studies are reported that confirm the anticommons thesis, which I think represents a further humiliation of that need of also respecting the "irrelevance" and the "implausibility" of guessed scientific hypotheses,[5] which is only occurring thanks to minimization of cognitive restrictions or obstructions.

A typical case of science privatized and commercialized is obviously represented by pharmaceutical industries. There is a widespread agreement among researchers that it is plausible to predict a further progressive decline of the pharmaceutical industries commitment to invest in Research and Development (R&D): new discoveries—and so new creative abductions—will dramatically decrease and, consequently, we will have fewer new drugs than expected. Currently, novel pharmaceutical discovery and development is already a very expensive, difficult, and inefficient process, especially due to the lack of models that accurately present the appropriate condition or that reflect the appropriate response. This means that ascertaining potential efficacy and off-target toxicity of drugs is a large part of the development process, often accomplished through the use of nonhuman animal models (Anson, Ma, & He, 2009).

Fewer targets and more whole-animal experimentation were replaced during the 1990s by more targets and in vitro screens. However, one key component of animal model-based reasoning is the admission that we do not know a lot of what we should know (or think we know): currently, several animal disease models are not predictive of clinical outcome (oncology, immunology, central nervous system (CNS) and other neurological and psychiatric conditions or pain—which are especially in need of new medications), whereas other models tend to be more reliable. We have to remember that protest against animal experimentation has had a heavy effect upon the pharmaceutical industry: unfortunately, the availability of well-characterized, reproducible, and moderate throughput animal models still furnishes a great reward during drug invention. The societal and regulatory consequences on animal experimentation have resulted in relocation to other countries, often involving obvious ethical outcomes. However, it is worth quoting the following interesting considerations provided by Bennani, concerning animal models predictivity: "Pharmaceutical research aimed at cancer treatment, neurological or psychiatric diseases, in particular, are areas in utmost need of new therapeutics, yet the most underserved from the discovery-to-clinical-success standpoint. As mentioned above, animal models in these disease areas are not predictive, yet regulatory agencies require preclinical investigational new drug (IND) packages to contain in vivo animal efficacy and data based on nonpredictive

[5] I have illustrated above in the first chapter of this book the role of irrelevance and implausibility as crucial to "productive" abductive discoveries.

or nonrelevant disease models. It is sad to see that many companies are giving up on these diseases rather than tackling the fundamental biological, pharmacological, and developmental causes of the failures. Perhaps the industry needs to tackle this very problem to demonstrably affect change in the success rate of oncolytics, neurological agents or neuroleptics, to name but a few" (Bennani, 2011, p. 784).[6]

The recent phenomenon of expensive drugs is already a clear consequence of the fact that pharmaceutical discovery and development is highly commercialized and at the same time a very expensive, difficult, and inefficient process. Indeed very expensive and very effective hepatitis C and cancer medications recently appeared in the market: private health insurers in U.S. and EU and welfare assistance in some EU countries limited prescriptions to few considered very sickest people. It seems we are facing with a kind of financial hazard—a kind of unintended, potential economic harm or paradoxical damage of therapy—which is consequently a critical issue that has to be seriously considered as some new treatments disregard the aspect of affordability to patients and their health systems.[7] Of course it can be said that insurers, by paying for the drug now, they can avoid later on paying for costly complications, for example liver transplants in the case of hepatitis C, also freeing up liver transplants for other people who need them. Is this true? Some economical modeling seems to confirm this prediction, but U.S. private insurance companies are skeptical in taking this long view, which involves a huge amount of short-term costs. Indeed they say that people regularly switch health plans, so insurers might not reap the long-term economic benefits of covering their members' hepatitis C treatments. Of course this fact discourages insurance companies from investing in hepatitis C treatments for more of their members: "The misalignment between short-term treatment costs and long-term benefits that private payers face, however, may not promote socially optimal treatment strategies" (Moreno et al., 2016, SP242).[8]

The rising costs of some drugs contract household budgets as well as federal and state budgets, Medicare, Medicaid, and other EU public health care programs that use taxpayer money to pay for drugs. The debate ranges from a discussion on how these expensive drugs affect the economy overall, for example by reducing the number of expensive hospital visits and allowing more people to get back to work faster. Anyway, in so far as we will not be able to solve the problem of actually having the people properly medicated using those expensive drugs, it will be difficult

[6] On some epistemological problems related to the role of animal models in bio-pharmaceutical research cf. for example (Shelley, 2006).

[7] The recent western financial crisis has increased the exposure of patients and health systems to the financial hazard, because of cuts in health budgets along with soaring unemployment and underemployment (Carrera & Olver, 2015).

[8] The case of expensive new drugs recently convinced few U.S. Politicians to hold some investigations into pharmaceutical companies such as Turing, whose former CEO Martin Shkreli said he would have raised the price of an anti-malarian drug by 5000%. Shkreli resigned after he was charged with securities fraud. Democrats have invited as a witness Valeant Pharmaceuticals executive Michael Pearson, formerly criticized for having increased the prices of two heart medications.

they will become cheaper. Not only, with the aging population, the health care costs and demands on price-control of medicines are expected to increase. We can also clearly note that the pharmaceutical industry's profitability and growth prospects are under pressure as health care budgets become increasingly strained because of the worldwide neoliberal policies which devastated the financial equilibria of modern western states.

The situation I have just described carries a clear consequence, which is exquisitely epistemological: we have to figure out how to address these problems, or run the risk that pharmaceutical companies will stop investing in the research and development of cures. Recently, in 2015, two researchers in the area of pharma consulting clearly addressed this aspect:

> [...] that cancer drug prices have been rising at rates well above the rate of inflation. This is of concern to all the major stakeholders because it could impact the quality of cancer care and bring undue financial hardship to the patient. In addition, it may compromise innovation and new drug discovery because it could force governments to implement arbitrary price controls. Government-mandated price drops have already occurred in some European countries. In addition, developing countries could also issue a compulsory license, which would enable local production of the patented drug. This is possible under the "Trade Related Intellectual Property Rights" agreement of the World Trade Organization and has already occurred with some HIV drugs. The Indian government did this for imatinib (Gleevec®), and it was upheld by the Indian Supreme Court. Such events do not support innovation and the discovery of new drugs to fight cancer (Dranitsaris & Papadopoulos, 2015, p. 299).

In summary, it is patent that both cognitive and technological creative processes in biopharmaceutical science,[9] which in the case of pharmaceutical companies are highly commodified, are (and will further be) seriously compromised. In these cases the commodification of science resorts to a form of objective "epistemic irresponsibility", which of course also has a quick practical outcome: the worsening of the general conditions of public health.[10]

---

[9] Already in 2010 (Paul et al., 2010) said "A key aspect of this problem is the decreasing number of truly innovative new medicines approved by the U.S. Food and Drug Administration (FDA) and other major regulatory bodies around the world over the past 5 years (in which 50% fewer new molecular entities (NMEs) were approved compared with the previous 5 years)".

[10] A recent research on the "commodification of science—often identified with commercialization, or the selling of expertise and research results—and the 'capitalization of knowledge'", as a threat to the autonomy of science and academic culture, also criticized for undermining the social responsibility of modern science, is richly illustrated in Radder (2019). Radder thinks that scientific knowledge constitutes a common good and challenges the legitimacy of commercial monopolies and the private appropriation and exploitation of research outcomes. Resnik (2021), in a recent review of the book in which he shows his basic agreement with Radder's treatment, also illustrates the typical counterarguments he thinks would have to be taken into account: (1) first, Radder disregards the important role that the patent offices and the courts play in defining and narrowing the scope of product patents; (2) getting rid of product patents would have a negative impact on private investment in research; (3) Radder also undervalues the important point that patents last at most 20 years, and that other researchers and companies can make and commercialize a product when a patent expires.

### 4.1.2 *Epistemic Irresponsibility II: How to Avoid the Eco-Cognitive Shutdown of Creative Abduction*

The reader would have noticed that the expression *epistemic irresponsibility* carries a moral halo which seems unexpected when associated to the word epistemology: let us further explain in detail my perspective on the interplay between epistemology and ethics and its relationship with my eco-cognitive approach. Stating that epistemology[11] and ethics are *entangled* does not only mean that reasoning and morality can be studied together, but rather that *it benefits* to study them together. The word *entanglement* is clearly borrowed from the language of quantum physics: even if the two philosophical disciplines have each their own theoretical dignity, many of the objects they deal with are just deeply entangled, so that ignoring one aspect or the other may cause a philosophical misperception of the matter at stake. For instance, by failing to appreciate the inferential dimension in a moral judgment and its enactment, or conversely, how moral priorities strongly inform and override our best hypothetical reasonings.[12]

The entanglement of epistemology and ethics has also tacitly emerged over the past recent years, transcending the philosophical impasses of the *is/ought* debate. Such appreciation seems to be more strongly nested in applied epistemology: Coady explicitly connects the origins of applied epistemology to the tradition of applied ethics (Coady, 2012, p. 1 and ff.), highlighting a theoretical practice of mutual borrowing that has characterized the different branches of philosophy since the very beginning. Clearing up the relationship, and the entanglement, between epistemology and ethics will help to shed light on another entangled relationship, that is the one between morality and violence. Indeed, the understanding of each theoretical entanglement (epistemology-ethics, morality-violence) rests on the understanding of the other, as the four poles are connected in a double dyadic system that I also explored in previous works (Magnani, 2011, 2015).

In this section I will deal with a second kind of epistemic irresponsibility that I think could be avoided taking advantage of some eco-cognitive suggestions that patently incorporate both methodological recommendations and also the commitment to a certain kind of what I call "moral epistemology",[13] such as in the case of the so-called "Innovation ASAP", I will soon illustrate.

[11] In this paragraph I exceptionally refer the word epistemology to the whole area of cognitive reasoned activities, such as it happens in a considerable part of current academic literature. Anyway, the reader would have realized—I also made a remark in the preface—that in the present book I adopted its classical intended meaning, which is only referred to scientific cognition.

[12] Some topics that powerfully display such entanglement are gossip studies (Bertolotti & Magnani, 2014), but also any epistemological approach on religion that cannot overlook how the violence entailed by religious cognition is rooted both in the moral assumptions and in the inferential regime that are typical of religion (Bertolotti, 2015), and overall the philosophical approach to the relationships between morality and violence (Magnani, 2011).

[13] I have quickly illustrated the concept of "moral epistemology" in Magnani (2017, chapter one).

**Table 4.1** Opportunities and pitfalls for big pharma R&D (Lundberg & Reilly, 2009, p. 441). ©2009 Elsevier, reprinted by permission

| Big Pharma R&D | |
|---|---|
| *Opportunities* | Pitfalls |
| Strong drug-hunting expertise. Accessing best scientists internally | Bureaucracy; process oriented static culture; poor corporate reputation |
| Genuine willingness to consult and learn from collective memory | |
| Broad technology platforms | Lack of focus with frequent strategic shifts |
| Value chain integration (Discovery, Development, Production, Market) | Unwillingness to change; 'silo mentality' |
| Global asset and knowledge exploitation, learning from failures | Not-invented-here' syndrome |
| Small unit feel (enpowerment) within large global organization (power) | R&D funding mainly late stage and short term |
| External transparent mindset | |
| Intelligent rist taking, broad portfolio | 'Me too' approach |
| Long-term innovation with constant improvements. Clear, sustained goals | Constant reorganizations and mergers—demotivation |

The crisis of pharmaceutical research in big pharma,[14] I have described in the previous section had been caused by reduced output of new medicines from research laboratories, drug pricing pressures, stricter regulatory environments, and the current economic downturn, together with the recent job losses and a considerable decrease of investments in Research and Development (R&D). This situation surely depicts the final result of previous eco-cognitive actions we see as informed by a remarkable epistemic irresponsibility, which surely harmed the productive performance of human abductive cognition in these areas of science and technology. Impoverished epistemic niches generated the reduction of scientific products and the breakdown of conceptual innovation.

Bennani (2011) first of all observes that previous investments in high throughput screening (HTS), combinatorial chemistry, genomics, and proteomics over the past two decades (the very technologies that were supposed to keep the industry from the abyss) have yielded a low return on investment, and consequently he suggests a complementary perspective on the corporate culture aspect of innovation. The epistemological aspects that are proposed are interesting and certainly highlight updated eco-cognitive good features needed to foster good creative abductions in science. The concept of "innovation ASAP" is introduced (an acronym for asking powerful questions, seeking the outliers, accepting defeat and populating astutely). In the background there is more than a suspect that the recent so-called

[14] A nice complementary summary of opportunities and pitfalls for big pharma Research and Development (R&D) is given in Table 4.1.

(mega) blockbuster business model has had a profound effect on the pharmaceutical industry, positive in terms of earnings but negative in terms of "epistemology":

> Nobody could argue, in this capitalistic world, against marketing a drug to achieve great revenues. Revenue enables further investment, with the goal of repeating an even greater achievement. Most major pharmaceutical companies have adopted this paradigm and invested heavily in R&D over the past two decades, more so than any other industry as a percentage of sales. This approach has forced R&D-intensive companies to focus mainly on opportunities with > US$ 1 billion annual market potential, which, in return, has created blind spots toward many diseases and perceived medical needs that did not meet these return levels. In addition, it tended to attract several players into similar markets, which led to many years of heavy investments to demonstrate clinical non-inferiority or marginal superiority to gain Food and Drug Administration (FDA) approval. The current trend to catch the current biologics or bio-similar(s) wave, could well backfire in a similar way to small-molecule follow-on approaches, over the next few decades. Furthermore, commercial functions have been in the driver's seat in clinical trial design, which has led to a business- rather than science-led clinical trial approach, often leading to failure. Although large revenues were sometimes achieved through this model, it seems to have led the industry to the edge of a cliff, given how difficult it has been to reproduce it routinely (Bennani, 2011, p. 780).

It is in this perspective of capitalistic competition, compelled to the innovation in drug discovery, that also a big pharma (Vertex) researcher and manager like Youssef L. Bennani has to recur to a program that is basically "epistemological", rather than *directly* "ethical", even seen as a "survival necessity": "innovation ASAP" (see above the various items that compose the acronym) is the motto, and the program seems to delineate a possible new fruitful policy able to foster creativity and technological innovation. Indeed the whole idea of a culture of innovation in the drug discovery settings remains unclear, given the multitude of pressure points: investment, competition, science knowledge, time, management, the diversity of sources of innovation and the still-unpredictable nature of medicinal chemistry, drug discovery and clinical outcomes: an approach which exemplifies some of the eco-cognitive features characterizing good creative abductions seems appropriate to solve the problem. Here the main aspects of the program that, as I observed above, incorporate at the same time both methodological recommendations and the commitment to a certain kind of "moral epistemology":

1. *asking powerful questions*. It echoes the specific character of creative abduction, which has to be performed in a knowledge-rich epistemic niche. The corporate epistemological environments have to avoid being dominated by a culture of copycats, in which all work on the same targets, all do combinatorial chemistry, all use outsourcing and the same consulting firms. All these negative aspects depict a strong limitation to that maximization of eco-cognitive openness that characterizes creative abduction in science;
2. *seeking the outliers*. "Outliers, as defined here, are a composite of key individuals, teams, observations, research approaches, work environment, corporate tolerance for, and reinforcement of, different perspectives, adequate reward systems, and the ability to seize those 'aha' moments: innovation is all about culture. Outliers are not necessarily individual contributors; they are observations, unique data relationships that do not fit the hypothesis or dogma, or that cause a

rethink of assumptions" (Bennani, 2011, p. 785).[15] Rules- or guidelines-based drug discovery is widespread in the pharmaceutical industries and represents a manager's dream: unfortunately the establishment of too many rules, both epistemological and pragmatic, constitutes barriers to innovation and to the finding of new answers. We have said that creative abductions are looking for guessed hypotheses where implausibility and irrelevance do not have to be discriminated: having rules at all levels of a human enterprise of science (cognitive activities included) is important and wise, but blindly following rules can also lead to "blindness" and to repress creativity and exceptional innovations;

3. *accepting defeat*. Looking for falsification is part and parcel of an epistemologically good scientific behavior. In the process of abductively guessing a new idea, one will need to accept failure. We know that in every process of abductive innovation the ability to deal with failure and to pass to new hypotheses is fundamental. The perseverance in the face of negative research outcomes is extremely negative but often affects R&D in pharmaceutical firms;
4. *populating astutely*. It suggests that, Bennani says, on the human resources side, a fundamental component of employing or forming a new team is related to a reasonable balance between people who disagree with the current standard of knowledge and those that execute orders. Organizations should promote people and teams able to enrich the various innovation incubators with complementary cognitive attitudes to foster creativity, rather than rewarding "me-too"[16] or "fast-follower" products.[17] Benchmarking itself can be creativity-threatening because it tends to only favor incrementally inventive products.

### 4.1.3 Epistemic Irresponsibility III: Neoliberalism Assaults to Epistemic Integrity of Biopharmaceutical Research

Interesting theoretical "philosophical" and sophisticated discussions about the ethics of pharmaceutical pricing could be further speculated, for example taking advantage of a Rawlsian view of distributive justice (Rawls, 1971; Spinello, 1992), contending that necessary drugs should be priced so that ability and willingness to pay will be within the reach of everyone, and a plenty of other argumentations

[15] Jaakko Hintikka would have called these outliers "oracles" (Hintikka, 2007), fundamental to perform good abductions (cf. section "The EC-Model of Abduction" in (Magnani, 2017, chapter one)).

[16] Copies of existing drugs, not exact copies because they need to be slightly different to be patentable.

[17] Brown reports the following amazing case related to the pathetic—and totally business-oriented—strategy of "inventing of a new disease": the drug maker Eli Lilly promoted a new medical disease concept, "premenstrual dysphoric disorder" (PMDD), allegedly in order to create new markets and patent protection for its drug Prozac (Brown, 2008, pp. 201–203).

could be advanced and further refined. The price of the AZT therapy puts it beyond the reach of a lot of people suffering from AIDS. Communicable diseases such as HIV/AIDS, malaria, and tuberculosis affect people that live mostly in third-world countries, receive only a small part of the investment per share in the global disease burden as compared to first-world diseases such as cardiovascular disease and diabetes. Gewertz & Amado (2004) obviously contend (and admonish) that "While the moral significance Rawls places on the basic liberty of life and the right to health is appealing and consistent with natural humanitarian concerns, the practical limitations associated with implementing these ideals are overwhelming. Ignoring economic factors and incentives in favor of providing treatment to every individual in need of medication, as Rawls's theory dictates, may undermine the ability of a company to function as a viable business" (Gewertz & Amado, 2004, p. 305). I think ethical lucubrations of this type, typical of the tradition of moral philosophy—even if very honorable—mean almost nothing in our neoliberal era, so I will immediately save my time and the time of the reader by renouncing to further propose these kinds of edifying but unfruitful—because a decent and appropriate context in which they could be listened to is lacking—intellectual exercises.

What is really the worst is that not only we are facing huge ethical problems, but also challenges to the epistemic integrity of modern science. Typical of scientific research in our era is the presence of strong relationships between universities and government researchers, on the one hand, and for-profit entities, on the other, which involve the funding of research projects by for-profit entities, consultancy arrangements, gifts, stock ownership, and management of start-up companies. In this context *epistemic integrity* of research is jeopardized, because scientists are biased toward the companies with whom they have relationships: conflicts of interest proliferate. The epistemologist James Robert Brown I have already quoted warns us about some epistemological perversions: "The editors [of the *New England Journal of Medicine* already in 2002] also claim in their editorial—and this is shocking—that it is increasingly difficult to find people to do reviews who do not have economic ties to the corporate world. If they left out such reviews, they claim, they would nothing at all on new products, leaving readers with no means of evaluation except that provided by the manufacturers themselves" (Brown, 2008, p. 194). Furthermore, by illustrating the famous Calabrese case (Shrader-Frechette, 2010) a list of four main ethical problems which afflict various fields of scientific research is provided (1) having financial conflicts–of-interest (extensive industry funding); (2) failing to follow guidelines for disclosing financial conflicts-of-interest; (3) failing to disclose coauthors' correct affiliations and conflicts-of-interest; and (4) censoring whistleblowers who reveal questionable behaviors. It is more than obvious that these four ethical problems can easily be responsible for damages to various levels of epistemic integrity.

Indeed, as I have already remarked in the previous subsection, the problem is that not only we are assisting to the birth of hyper expensive drugs (for example the recent cases of Hepatitis C and some cancer treatments) but the current economical system challenges the chances of making new discoveries themselves, and so the possibility of freely exercising human creativity and capacity to discover and innovate. It can be clearly said that economic medicalization of research "[...] is a process where the traditional values associated with the scientific method are replaced by research ethics that reflect the values of the market place" (Poitras & Meredith, 2009, p. 318).

Reiss (2010, pp. 435–436) usefully remembers to us that care legislation in the United States since 1980 has been extremely "business-friendly", not always acting with the patient's welfare in mind and "doing violence to the production of scientific knowledge". Inadequate regulation ("both the failure to regulate certain practices as well as the introduction of new regulations with adverse effects") has created a completely business-oriented system where everyone can make profits: "companies and their stockholders by high sales and profit margins; doctors, by gifts and financial contributions from the industry; academic researchers, by the possibility to patent the fruits of their research; politicians, by board memberships in the industry and campaign donations; regulators, by so-called 'user-fees', paid for by the industry; and 'apparently' patients, by an allegedly highly innovative industry that is spending billions to find new treatments for their health problems".

In our era the practice of neoliberal economics is strongly associated with a "one dimensional" ethics that privileges market valuation principles over all others, and, regrettably, the epistemological ones included. How can the prevalence of the idea of "economic liberty", when privileged over all others, in the context of a deteriorated democratic processes, jeopardize, and damage or pervert received epistemic values? In the previous subsections I have described some of the so-called "impoverished epistemic niches", which represent an attack to the flourishing development of human creative abduction: it is obvious that at the basis of such a challenge we can see a system that putting profits over people will likely lead to a shortage of essential drugs and instead to an increment of irrational, unessential, disused, and insecure drugs, with thousands of brand names, promotional vacuities, etc. After all big pharmaceutical corporations pursue strategies consistent with shareholder wealth maximization (SWM).

Neoliberalism plus globalization also lead to more experimentation of new drugs to third world countries and often the subjects of these trials are not be clearly given the information about risks:[18] the main strategic decision variables at stake are pricing, quantity to be produced, advertising and publicity, product—often superficial—reshaping and development, and investment as strategic decision variables. Pricing may be largely determined by marketing considerations, of course related to compensating costs by suitable margin, which would be decided by contribution to fixed costs and the profit rate and profitability commanded by the capital markets (D'Mello, 2002, p. 180). In sum, being a product and process innovations are part of a competitive strategy and have to be coherent with advertising and promotion: "This may entail deliberate product proliferation, product

[18] In the western world doctors are sometimes hugely paid to favor these experiments so opening the door to possible abuses (Brown, 2008, p. 195).

differentiation, including irrational combinations, the manipulation of patents and the codes and standards set voluntarily by an industry".

Moreover, medicalization is a social process where the medical profession involves not only doctors and associations of doctors, but also the pharmaceutical industry providing the drugs, the academic institutions and journals involved in forming doctors and promoting fundamental research activities, the government granting agencies and other sponsors, etc. As for now, given the fact that the source of capital for the pharmaceutical industry is the global financial markets, the primary ethical motivations of big pharma profoundly differ from those of the other actors.[19] A dilemma for government regulators arises: how to balance public health needs "with the need to restrict the economic footprint of the regulatory framework on an industry that produces and distributes some of the most important products of modern science?" (Poitras & Meredith, 2009, p. 313). After all since the famous public policy disaster generated by thalidomide in the 1960s, it has been acknowledged that in medical research and development (R&D) the conflict of interests between profit-maximizing of private sphere firms and those of governments would need severe regulatory supervision.

Another aspect of current business-oriented medicalization is related to the fact that it is becoming a process where more and more aspects of everyday life come under medical dominion, influence and supervision. Poitras & Meredith (2009, p. 315) also stress the excesses of the so-called "marketing of disease" and illustrate a "list of categories consistent with economic medicalization, including: senior male disorders, such as andropause, baldness, and erectile dysfunction; behavioral disorders, such as hyperactivity in children and adult ADHD; and, biomedical enhancements, such as human growth hormone and steroids".

Finally, here is a list of some derived forms of commercialization or commodification of aspects of the biomedical research (also other aspects are included), which the neoliberal era has transformed into real challenges to epistemic integrity, which the philosopher of science has the duty to highlight:

1. instead of trying to reduce the possibility of bias in the interpretation of the observed data, the "moral codes" of the marketplace "require firms to be basically concerned with abnormal gains (losses) associated with 'positive' (negative) research results": in few words, economic medicalization "occurs because there is a decided bias towards unjust acceptance and against unjust rejection" (Poitras & Meredith, 2009, p. 318). The tendency is to publish only favorable clinical trial results regarding an experimental drug and suppress discordant evidence, avoiding any kind of adversarial system (Biddle, 2013). It is obvious that it is not

[19] The epistemologist Brown presents a very negative description of big pharma: "The swaggering entrepreneurs of the pharmaceutical industry boast that they are doing risky, innovative research. This is pure nonsense. [...] Big pharma consists of business people exploiting the intellectual work of others. Frankly, who needs these parasites? Their only contribution to medicine has been marketing ..." (Brown, 2008, p. 205).

rewarding, so to say, to present negative information concerning products, which already costed millions of dollars of investments;[20]

2. the so-called "muzzle clauses" in the contracts—which can involve lawsuits etc.—intend to prevent researchers from releasing any information about the clinical trial without the sponsor's permission: when negative data are found they tend to remain secret, even when could predict dangers to public health. The knowledge in motion which I have described as necessary in Sect. 3.5.1 above is stopped: scientific information cannot appropriately circulate. But also terrible pragmatic and not only epistemic costs and consequences can derive. The tragic consequences of OcyContin, Neurontin, Paxil, Accutane, Baycol (Caplovitz, 2006), Aprotinin and Vioxx (Avorn, 2006)[21] testify more or less terrible dangers and speak clearly regarding long-term effects emerged only after long periods of time already in the market place, due to a lack of preventive trials before marketing;
3. opinion leaders (also academics) and companies are too often complicit: for example they are financed for lectures at conferences or act as paid consultants to some corporations so showing low degrees of independence. Some drug companies adopt the so-called "ghost-writing" when they send the results (only the positive ones) of their research to an author trying to secure her endorsement to favor acceptance of their new drugs in the marketplace. Academics are often seduced by the chance offered of publishing in prestigious journals appropriate to favor career progress. Symmetrically, it is unlikely that unsuccessful clinical trials will be published and physicians are in general exposed to articles that support certain drugs or devices;
4. commercialization can affect even doctors' diagnostic inferences (selective abductions, as we know). The reliance of drug detailing and sampling on pure marketing tactics raises suspicion of unethical economic medicalization. Drug and medical devices are offered and sales representatives bring research literature and clinical trial results to the doctors in efforts to influence their prescription pattern choice (Turner, Matthews, Linardatos, Tell, & Rosenthal, 2008). Representatives make donations of free samples to hospital and doctors

[20] Elliott links this behavior to the concept of the so-called *selective* ignorance: "[. . . ] the category of selective ignorance can sometimes merge with the category of ignorance as a strategic ploy or active construct. For example, industry groups may purposely study the beneficial or neutral effects of their products while avoiding research that might yield negative information" (Elliott, 2012b, p. 332).

[21] Justin Biddle reports findings to the effect that Merck, the producer of Vioxx, mischaracterized the state of knowledge regarding its possible cardiovascular side effects between 2000 and 2004 and accounted for data on the drug deficiently (Biddle, 2007). 17 members (against the remaining 15) of the FDA advisory committee in February 2005 decided to put Vioxx back on the market, after Merck had removed it, and 10 of those 17 had financial conflicts of interest with at least one of the pharmaceutical companies that had drugs under examination, such as Merck, Pfizer, and Novartis (Biddle, 2013). It is more than obvious that regulatory agencies have a very small influence on private companies, given the fact such agencies are subjected to corporate funding and to committee members who are scientists that are related to those companies.

tending to insidiously influence general medical staff, avoiding firstly contact pharmacologists, who are more expert and so more able to criticizing and evaluate drugs;

5. clinical trials themselves[22] are more oriented by marketing than by deep R&D motivations: the physicians can licitly receive fees for the recruitment and tracking of subjects admitted into the clinical trials. Two particularly famous cases are (1) Bayer admitting to a "mistake" in suppressing a study that showed dangerous side-effects related to the drug Baycol, and (2) the already cited Merck's suppression of research that showed Vioxx doubled the risk of myocardial infarction and stroke (Avorn, 2006). Only the fact these drugs presented adverse outcomes in a large quantity of treated people made possible the discovery of these misconducts, which are horrible certainly from the ethical point of view but in our perspective, especially from the epistemological point of view.

In sum, epistemic decency and integrity are constantly in danger or at risk thanks to the neoliberal attack—through excessive, almost anarchistic, commercialization—to academy and public research, but we have to say that also other more practical aspects were not favored by the neoliberal turn, such as for example the ones described by Brown in 2008 in a famous article published in *Science*:

> The United States is unique among industrialized countries in not having a national health system. Health care is overwhelmingly private and largely in the hands of insurance companies. The cost is approaching 15% of the U.S. gross domestic product and more than one-quarter of the population is not covered. By contrast, Canada (like most other industrialized countries) has universal coverage at a cost of under 9% of gross domestic product. Aside from the cost, it is hard to compare the relative quality of the health-care systems, but one statistic is revealing: Cancer patients in Canada live an average of 14 months longer from the time of detection than those in the United States (Brown, 2000, p. 1702).[23]

We academics all know that philosophers of science are not involved—neither directly nor indirectly—in decisions regarding public policy: it seems that policy decisions must be based on social goals and other factors that have nothing to do with epistemology. At this point it is clear that various aspects of current decision making in commercialized research are jeopardizing scientific creativity and human creative abduction in general. As an epistemologist I am just indicating the problem: too many aspects of the socio-economical organization of science—increasingly business-oriented—put in danger the genuine tradition of objective science and

[22] The outsourcing of medical research has become a strong imperative in the global pharmaceutical industry. Pressed by competition, the importance of speed in drug development, and higher domestic costs, pharmaceutical companies have to outsource critical contracts, clinical research, and drug testing, in general into emerging markets.

[23] See also the Wikipedia entry: https://en.wikipedia.org/wiki/Comparison_of_the_healthcare_systems_in_Canada_and_the_United_States.

some of the best aspects of the human abductive scientific cognition.[24] I do not have practical suggestions and solutions, but surely the smart people who are involved in economical and political decisions and at the same time are interested in favoring the further growth of science in the western world will find ones.

## 4.2 Optimizing the Eco-Cognitive Situatedness: Human Creative Abduction between Academia and Corporations

To further analyze the role of the optimization of eco-cognitive situatedness in its relationship with creative abduction in science it is important to deal with some interesting aspects of the current interplay between academia and corporations. The case of biopharmaceutical research is still exemplar. Still now, in an era of powerful attack on higher education (Giroux, 2014), innovation and creativity are generally fostered in academia, but pharmaceutical firms tend to redefine the word "new" as a trivial "me-too" or "me-better" (just modifications of existing substances), as I have illustrated in the Sect. 4.1.2 above. Indeed biomedical research basically produce drugs that are not at all innovative—but often very expensive—with uncertain medical value and often not clearly checked from the point of view of safety and effectiveness.

These scarce innovative fruits also derived from a change of the nature of academia received from the past. For example, in the case of biopharmaceutical research, profit-oriented companies have substituted academic and other non-profit institutions in performing various medical research such as for example clinical trials. Reiss observes that 71% of industry-sponsored clinical trials were outsourced to academic medical centers in 1991 but, by 2001 this percentage had dropped to 36%; moreover, we have already said that industry-sponsored research tends to present results that are favorable for the tested drug. The commercialization of research has also attenuated the standards and drugs are sometimes tested on patients who are younger and more healthy than the general target population. Finally, a considerable number of clinical trials is never published or described in a way that makes possible positive judgments about the validity of the reported results (Reiss, 2010, p. 431–432). Indeed, some of the countries targets of medical trials outsourcing have been seen as serving of "guinea pigs" to the world (D'Mello, 2002; Shah, 1993), and also clearly testify the shift of clinical research and drug development from academy to commercial contract research organizations. Multinational companies like Pfizer, Eli Lilly, GlaxoSmithKline, Sanofi Aventis, and Roche have outsourced clinical studies abroad, especially in India.[25]

[24] Indeed it seems that other political, economical, and social processes are jeopardizing the good performance of various human expert cognitive skills, not only creativity. On social networks as ignorance-makers cf. (Arfini, Bertolotti, & Magnani, 2019).

[25] Various cases of clinical trial misconduct are described in Adobor (2012).

Academic research remains relevant in commercially unattractive therapeutic areas, such as rare disorders and parasitic diseases, and in research into natural products, which are robustly disregarded by the big pharma.[26] Unfortunately, after the 1990s, the golden era of big pharma, the productivity is now very low and the R&D costs are rising, and the dissipation of proprietary products and "dwindling pipelines" are still active (Khanna, 2012).[27] In the case of biomedical research, we are currently in front of a kind of unsurprising paradox: in academy great biomedical science is frequently done, but without immediate applicability to human pathologies, instead pharmaceutical firms are engaged in a kind of "drug hunting" often far from being highly innovative but simply resorting to an activity of making medicines that just work in human beings. It is well-known that genomic sequencing will not deliver large numbers of new, valid targets: the anti-infective, antiviral and antiparasitic areas have had the complete genomic sequence of their target organisms available for many years. Similarly, understanding the etiology of a disease is not necessary positively linked to finding treatments: cystic fibrosis is well studied and understood, but is still poorly treated.

Bennani eloquently observes that the different organizational structure of industry of research might be responsible for the failures and of the low level of innovative results:

> [...] higher education systems are organized around scientific disciplines and departments, whereas researchers naturally tend to congregate around similar models in the industrial setting. Alternatively, the industry tends to be organized around diseases or therapeutic areas, with emphasis on centers of excellence. There are few organizational models based around "drug discovery excellence" or "integrated disease discovery and development centers" or, better still, "chemical biology-pharmacology", "pharmaceutics development", and so on. The point here is that much innovation, speed, productivity are probably lost through lack of better integrative organizational, cultural, and scientific communication models. Innovative organizational and management models, along the entire spectrum of R&D, will be key in helping address current shortcomings (Bennani, 2011, pp. 784–785).

The contraction in research within pharma is surely due to the lack of that eco-cognitive activity in which an environment characterized by that "knowledge in motion" I have described above in Sect. 3.5.1 is active. This decline seems to be accompanied by a renaissance of research within the academic setting: often, groups grow organically from academic research laboratories, exploiting a particular area of

---

[26] Big pharma are generalists and orphan drug development represents a small proportion of their product portfolio (Bruyaka, Zeitzmann, Chalamon, Wokutch, & Thakur, 2013). I have already said in Sect. 4.1.1 that the excess of patenting tends to discourage the creation of new drugs, and especially for the third and developing world. In these worlds medicines cannot be easily found: a commodified scientific research privileges the medicines that already exist and, moreover, the program of developing new drugs to treat diseases that afflict primarily or exclusively the developing world is fundamentally disregarded.

[27] In between 2008 and 2012, the top ten pharma companies have lost more than 200,000 jobs. A decline has not stopped even considering the exploitation of both China and India, which are offering less expensive preclinical and clinical services (at savings of 30–80%), in a situation in which informed consent and patient safeguard are often—de facto—lacking (Lee Chang, 2016).

novel biology or new technology and we are facing an increment of drug discovery activities in the academic, charitable, and non-forprofit sectors. This is the case of the Cancer Research UK (CR-UK) where the drug discovery team at the Manchester Institute has been established to translate novel research from the Manchester cancer research community into drug discovery programs (Jordan, Waddell, & Ogilvie, 2015).

Anyway, the general comments concerning the future of drug discovery are rather pessimistic, as eloquently illustrated by the following passage, published in the relatively recent and rich collective book *Drug Discovery and Development*:

> However, there is cause for concern. Declining government funding and reformed educational policies in the western world are likely to have serious implications for drug discovery educators and practitioners, which could widen the already significant gap between research scientists at the highest level and the education of students at undergraduate and postgraduate levels. There is a real concern that the scientists of tomorrow will not possess the "right", tools in the toolkit to be able to effectively interrogate and address the questions being asked by research scientists in academia and industry today. The challenges can only be met if the government agencies worldwide are willing to invest in the education of academics and students alike (Pors, 2011, p. 91).

### *4.2.1 "The Symbiotic Model of Innovation" and the Precompetitive Collaborations*

Crisis and devastating policies are already leading pharma explore external alliances to further stimulate the stagnating innovative areas. This seems to offer a new opportunity for collaboration between academic institutes and industry. Centuries of academic scientific tradition has contributed gigantically to developing better knowledge of hundred of disease states and to the respective therapies: of course this tradition has always been less committed to the practical aim of creating new drugs and at the same time not particularly sensitive to intellectual property protection. During the last 20 years the strong reduction of funds devoted to academic research from government and private agencies has created new attention of university researchers to industry for support in fundamental research and identifying areas of mutual interest. Steve Carney, editor of the international journal *Drug Discovery Today* thinks that the "[. . . ] days when big Pharma companies operated in glorious isolation, finding their own targets and performing medicinal chemistry in house to find that elusive blockbuster compound are, generally, over" (Carney, 2016).

In the previous sections I have amply illustrated that, in a commercialized model of research, creativity is jeopardized certainly because innovation chains and knowledge in motion are humiliated. Given the simple fact that every new idea is built on a previous idea, if the previous idea is protected, the new idea cannot be developed.[28] In presence of an excess of patents it is clear that in some areas of science, continuing research and trying innovation can be guaranteed only by some patent infringement. For example, patents in areas such as DNA diagnostics seem to jeopardize the development of certain kinds of knowledge and certain technological capabilities that have the potential to save lives (Biddle, 2012). Patents and property rights constitute a serious obstacle to perform research (both "normal"—in Kuhnian sense—and innovative) and also to disseminate research results. Recent patent infringement cases in the U.S., such as Monsanto vs. McFarling (2004), Monsanto vs. Scruggs (2006), and Monsanto vs. Bowman (2011) are described in (Biddle, 2014):[29] Monsanto prohibition of seed saving seems to violate the doctrine of patent exhaustion, which states that a patent holder's right to exclude others from using or selling an invention is exhausted after an authorized sale.

Another threat to the epistemic integrity of research can be easily found in the field of ethnopharmacology: astonishingly we see in danger the celebrated rational criterium of falsification, jeopardized by the pressure to publish that afflicts those scientists. In the interpretation of molecular pharmacological data published in ethnopharmacology-related journals a very unsatisfactory level of self-evaluation is observed. The problem is that there has been a rebirth in scientific study of medicinal plants with medical effects, also enriched with new methods for the exploration of new drugs candidates for several diseases. We academics all know that various new policies regarding the evaluation of researchers have been implemented in the last years with the aim of improving research and its "productivity" and the "merit" of people. Consequently scientists are in general under pressure to publish. In some cases—ethnopharmacology is just a minor example—this fact weakens the commitment to falsify the theories, so generating epistemic irresponsibility.

Gertsch reminds us that we may obviously ascertain when a hypothesis or theory is wrong when it can be "proved" wrong (Popper, 1959, 1963; Popper & Miller, 1983) and observes

> In an "inflation-driven" scientific environment the currency (scientific publications) is devaluated as the amount of currency increases, which motivates us to publish as many claims as possible. [...] The result is the fabrication of unscientific hypotheses and theories. Is it correct to talk about an antimalarial activity of an extract when the readout is derived from an in vitro model? Is a compound an anticancer agent only because it kills cancer cells in vitro? If compound *X* shows anticancer effects in different in vitro models but is completely inactive in vivo because it is not bioavailable or metabolized we have the possibility of improving its chemical properties such that it becomes a therapeutic candidate. [...] Only when researchers start to try to falsify or refute their observations and subsequently announce grown-up theories will we see an improvement in scientific quality and a more efficacious promotion of knowledge. In my opinion this is urgently needed in ethnopharmacology (Gertsch, 2009, pp. 181 and 182).

---

[28] I have described this issue above in Sect. 4.1.1 (cf. p. 81), when dealing with the so-called "Tragedy of Anticommons".

[29] In (Biddle, 2014) many other examples are illustrated and some possible proposals for overcoming this situation of epistemic hazard advanced: voluntary agreements to make research feasible, elimination of patents, and research exemption.

Even if the situation of commercialized biopharmaceutical research (and agricultural research, we have to add) presents all the epistemological problems that I have illustrated in this chapter, fortunately, there are still alive many areas in which commercialization of science is relatively not present and scientific progress very relevant, for example in mathematics, high energy physics, and evolutionary biology, which are patent-free.[30] Of course these areas are considerably damaged in their intellectual freedom because of a chronic lack of funds and especially because of the obsessive request—addressed to all the academic departments—to organize their courses and their research activities primarily for forging "workers", to be promptly used by the outside market.

To re-create the eco-cognitive conditions for favoring creative scientific abduction pharmaceutical companies have to look outside of their towers with the aim of generating new ideas, identifying targets, and reducing redundancy. In this process academy will still play a fundamental role: sharing information early in the scientific cognitive processes will enhance it without compromising the "sacred" intellectual property positions.[31] Some articles recently published in *Drug Discovery Today* have just highlight some of these areas and the benefits of the so-called *precompetitive approaches* or *collaboration*. I have to say that it is really amazing to see invented new neologisms exploiting the adjective "precompetitive" to describe the creation of partnering models comprising both academic departments and big and small pharma industry. These unions seem appropriate to develop a better understanding of complicated diseases, identify and validate new therapies, seek and validate novel biomarkers for prediction of efficacy and safety. I was saying that it is amazing to see the adoption of the neologism "precompetitive": it is also very revealing, is not it the admission of a strong failure of the attack to higher education and academic scientific research with its paranoic emphasis on commercialization and competition? The many decision-makers who follow neoliberalism are just discovering—I have to say very slowly—that, having killed a great part of the state structures and private financial funds reserved to universities, by concentrating everything in the private sector thanks to the adoption of a totalitarian business-oriented model, they have also created and they are still—obstinately—developing an eco-cognitive environment for scientific research in which innovation, creativity, and so good human creative abductive cognition are highly jeopardized.

[30] In these cases Brown observes that "Curiosity, good salaries, and peer recognition are motivation enough" (Brown, 2008, p. 209).

[31] An interesting treatment of various problems related to intellectual property in life sciences taking advantage of the history of the international patent regime and the life science industries, both of which traced back to the late nineteenth century, is offered by Dutfield (2009).

The partners of these precompetitive unions return to have open access to the data, technology, and training options to utilize in their own research or proprietary projects. Already in 2012 Ish Khanna, experienced pharmaceutical researcher and manager, said that

> With constrained microeconomics, new creative deal structures and alliances are emerging with emphasis on partnering to share risk and rewards. Although innovation is a driver in deal-making, interest in early collaborations seems to have intensified during the past two to 3 years. This is a marked shift from previous models wherein biotechs would carry projects to clinical proof-of-concept before partnering with larger companies. Early partnerships, equity holdings, option alliances with non-dilutive funding are emerging as desirable alternates. The risk shared deals (preclinical and clinical) with preferred partners are becoming common place (Khanna, 2012, p. 1098).

The so-called "Symbiotic Innovation Model" is related to this new research perspective in terms of early partnership and aims both at benefiting all partners and at being particularly relevant for the so-called novel mechanism-based drugs. It seems that variations of this model will likely be adopted rapidly in pharma industry during coming years.[32]

Reopening the barriers between public and private biopharmacological sectors thanks to the precompetitive collaborations is a first step to favor the "creative scientist", that is the unconventional thinker who currently feels imprisoned under constrictive environment. Human creative abduction can flourish beyond limits only in eco-cognitive open systems.

It is simple to detect and criticize behaviors that testify patent forms of epistemic irresponsibility, for example a corporation that refuses to reveal information about the harmful effects of its products, but it will be more difficult to criticize studies organized in such a way of minimizing the chance of detecting negative effects. It is obvious that selective ignorance is always at stake, and this regards the directions of research science has to adopt, and that often reflect the interest of restricted groups:[33] criticizing it is related to the adoption of a different interest, for example the one of the public goods. Another count is the critique to the threats to the

[32] One such approach, the European Lead Factory (ELF), a project of the Innovative Medicine Initiative, has recently created a collaborative lead generation platform to share research in the early moments of drug discovery. "The state-of-the-art high-throughput screening (HTS) infrastructure and the industrial-quality Joint European Compound Collection of the ELF are made available at no cost to European research investigators, with a milestone payment system applied to any exploitation projects targeting commercialization (https://www.europeanleadfactory.eu/)" (Karawajczyk et al., 2015, p. 1310). The European Lead Factory is a collaborative public-private partnership aiming to deliver innovative drug discovery starting points. Having established the first European Compound Library and the first European Screening Centre, the EU Lead Factory gives free access to up to 500,000 novel compounds, a unique industry-standard HTS platform, and much more.

[33] Sometimes it is the mere neoliberal organization of universities and research centers which inhibits freedom and that eco-cognitive openness and optimization of eco-cognitive situatedness I have introduced in this book. For example: idiot pressure to produce science that can be published in disciplinary important journals, huge bureaucratization of academic departments that inhibits creative collaborations and, so to speak, both "enthusiasm for understanding" and needed times for

productivity of science itself I have illustrated in the previous sections, which jeopardize the entire way of producing rational knowledge.

In conclusion, political and academic strategies can be implemented to restore epistemic responsibility when needed (for example (Elliott, 2012b) quotes political activism and deliberative forums for both steering science toward public goods and identifying public goods thanks to democratic policies). The success of these processes will exclusively depend on the decisions and actions the human beings will adopt (or will be able to adopt).

## Summary

In this chapter the problem of discoverability has been characterized by the analysis of some current difficulties that affect various aspects of the scientific enterprise such as in the case of the organization of Research and Development (R&D) in biopharmaceutical companies. I contended that this case symbolizes a paradigmatic example of what I have called "impoverished epistemic niches". It seems that some of the fundamental aspects that characterize modern science are jeopardized: in the light of the importance, I repeatedly stressed in the previous chapters, of the crucial role played in scientific abductive cognition by the so-called "maximization of eco-cognitive openness and optimization of eco-cognitive situadedness", I have examined various types of what we can call "*epistemic irresponsibility*". I have illustrated the current increasing expansion of commodification and commercialization of science, marketing of technoscientific products, impoverishment of the epistemic niches, showing that a consequence is and could further be an attack to the possibility of a flourishing of human fecund and successful abductive creative cognition in science, and so at the same time human creativity severely compromised.

## References

Adobor, H. (2012). Ethical issues in outsourcing: The case of contract medical research and the global pharmaceutical industry. *Journal of Business Ethics, 105*(2), 239–255.

Anson, B., Ma, J., & He, J. (2009). Identifying cardiotoxic compounds human iPS cell-derived cardiomyocytes (and other cell types) streamline quest for novel drug candidates. *Genetic Engineering and Biotechnology News, 29*(9), online.

Arfini, S., Bertolotti, T., & Magnani, L. (2019). Online communities as virtual cognitive niches. *Synthese, 196*(1), 377–397.

Avorn, J. (2006). "Dangerous deception": Hiding the evidence of adverse drug effects. *The New England Journal of Medicine, 355*, 2169–2171.

"concentration", lack of funds, funds that compel to study certain themes, obsession for increasing profits, etc.

Bennani, Y. L. (2011). Drug discovery in the next decade: Innovation needed ASAP. *Drug Discovery Today, 16*(17/18), 780–792.

Bertolotti, T. (2015). *Patterns of rationality: Recurring inferences in science, social cognition and religious thinking*. Heidelberg/Berlin: Springer.

Bertolotti, T., & Magnani, L. (2014). An epistemological analysis of gossip and gossip-based knowledge. *Synthese, 191*, 4037–4067.

Biddle, J. B. (2007). Lessons from the Vioxx debacle: What the privatization of science can teach us about social epistemology. *Social Epistemology, 12*(1), 21–39.

Biddle, J. B. (2012). Tragedy of the anticommons? Intellectual property and the sharing of scientific information. *Philosophy of Science, 79*(5), 821–832.

Biddle, J. B. (2013). Institutionalizing dissent: A proposal for an adversarial system of pharmaceutical research. *Kennedy Institute of Ethics Journal*, *23*(4), 325–353.

Biddle, J. B. (2014). Can patents prohibit research? On the social epistemology of patenting and licensing in science. *Studies in History and Philosophy of Science Part A, 45*(1), 14–23.

Brown, J. R. (2000). Privatizing the university–the New Tragedy of the Commons. *Science, 290*(5497), 1701–1702.

Brown, J. R. (2008). The community of science®. In M. Carrier, D. Howard, & J. A. Kourany (Eds.), *The challenge of the social and the pressure of practice: Science and values revisited*. Pittsburgh: University of Pittsburgh Press.

Bruyaka, O., Zeitzmann, H. K., Chalamon, I., Wokutch, R. E., & Thakur, P. (2013). Strategic corporate social responsibility and orphan drug development: Insights from the US and the EU biopharmaceutical industry. *Journal of Business Ethics, 117*(1), 45–65.

Caplovitz, A. (2006). *Turning medicine into snake oil: How pharmaceutical marketers put patients at risk*. Trenton, NJ: NJPIRG—New Jersey Public Interest Research Group Law and Policy Center.

Carney, S. (2016). The topic of this month's newsletter from drug discovery today is "precompetitive intelligence". *Drug Discovery Today.com*, 09 May, online.

Carrera, M., & Olver, I. (2015). The financial hazard of personalized medicine and supportive care. *Supportive Care in Cancer, 23*(12), 3399–3401.

Coady, D. (2012). *What to believe now: Applying epistemology to contemporary issues*. New York: Blackwell.

D'Mello, B. (2002). Transnational pharmaceutical corporations and neo-liberal business ethics in India. *Journal of Business Ethics, 36*, 165–185.

Dranitsaris, G., & Papadopoulos, G. (2015). Health technology assessment of cancer drugs in Canada, the United Kingdom and Australia: Should the United States take notice? *Applied Health Economics and Health Policy, 13*(3), 291–302.

Dutfield, G. (2009). *Intellectual property rights and the life science industries a twentieth century history* (2nd ed.). Singapore: World Scientific.

Elliott, K. (2012a). *Ignorance, uncertainty, and the development of scientific language*. In N. Janich, A. Nordmann, & L. Schebek (Eds.), *Nichtwissenskommunikation in den Wissenschaften. Interdisziplinäre Zugänge*, (pp. 295–316). Frankfurt am Mein, Peter Lang.

Elliott, K. (2012b). Selective ignorance and agricultural research. *Science, Technology, and Human Values, 38*(3), 328–350.

Gertsch, J. (2009). How scientific is the science in ethnopharmacology? Historical perspectives and epistemological problems. *Journal of Ethnopharmacology, 122*(2), 177–183.

Gewertz, N. M., & Amado, R. (2004). Intellectual property and the pharmaceutical industry: A moral crossroads between health and property. *Journal of Business Ethics, 55*(3), 295–308.

Giroux, H. A. (2014). *Neoliberalism's war on higher education*. Chicago, IL: Haymarket Books.

Heller, M. A., & Eisenberg, R. S. (1998). Can patents deter innovation? The anticommons in biomedical research. *Science, 280*(5364), 698–701.

Hintikka, J. (2007). *Socratic epistemology: Explorations of knowledge-seeking by questioning*. Cambridge: Cambridge University Press.

Jordan, A. M., Waddell, I. D., & Ogilvie, D. J. (2015). Rethinking "academic" drug discovery: The Manchester Institute perspective. *Drug Discovery Today, 20*(5), 525–535.

Karawajczyk, A., Giordanetto, F., Benningshof, J., Hamza, D., Kalliokoski, T., Pouwer K., . . . Tzalis, D. (2015). Expansion of chemical space for collaborative lead generation and drug discovery: The European Lead Factory Perspective. *Drug Discovery Today, 20*(11), 1310–1316.

Khanna, I. (2012). Drug discovery in pharmaceutical industry: Productivity challenges and trends. *Drug Discovery Today, 17*(19/20), 1089–1102.

Lee Chang, P. (2016). The abandoned stakeholders: Pharmaceutical companies and research par ticipants. *Journal of Business Ethics, 143*(4), 721–731.

Lundberg, J. M., & Reilly, C. (2009). The road ahead for large pharma: Long-term science and innovation. *Drug Discovery Today, 14*(9/10), 439–441.

Magnani, L. (2011). *Understanding violence. The intertwining of morality, religion, and violence: A philosophical stance*. Heidelberg/Berlin: Springer.

Magnani, L. (2015). Violence and abductive cognition. Epistemology and ethics entangled. In L. Magnani, P. Li, & W. Park (Eds.), *Philosophy and cognitive science II: Western and eastern studies* (pp. 95–113). Heidelberg: Springer.

Magnani, L. (2017). *The abductive structure of scientific creativity: An essay on the ecology of cognition*. Cham, Switzerland: Springer.

Moreno, G. A., Mulligan, K., Huber, C., Linthicum, M. T., Dreyfus, D., Juday, T., . . . Lakdawalla, D. N. (2016). Costs and spillover effects of private insurers' coverage of Hepatitis C treatment. *The American Journal of Managed Care, 22*(5), SP236–SP244. Special Issue: HCV, No. 6, published online.

Paul, S. M., Mytelka, D. S., Dunwiddie, C. T., Persinger, C. C., Munos, B. H., Lindborg, S. R., & Schacht, A. L. (2010). How to improve R&D productivity: The pharmaceutical industry's grand challenge. *Nature Reviews. Drug Discovery, 9*(3), 203–214.

Poitras, G., & Meredith, L. (2009). Ethical transparency and economic medicalization. *Journal of Business Ethics, 86*, 313–325.

Popper, K. R. (1959). *The logic of scientific discovery*. London, New York: Hutchinson.

Popper, K. R. (1963). *Conjectures and refutations: The growth of scientific knowledge*. London: Routledge and Kegan Paul.

Popper, K. R., & Miller, D. (1983). A proof of the impossibility of inductive probability. *Nature, 302*, 687–688.

Pors, K. (2011). Drug Discovery into the 21st Century. In I. M. Kapetanovic (Ed.), *Drug discovery and development—present and future* (pp. 69–96). Rijeka, Croatia: InTech.

Proctor, R. N., & Schiebinger, L. (Eds.). (2008). *Agnotology: The making and unmaking of ignorance*. Stanford: Stanford University Press.

Radder, H. (2019). *From commodification to the common good: Reconstructing science, technology, and society*. Pittsburgh: University of Pittsburgh Press.

Rawls, J. (1971). *A theory of justice*. Cambridge, MA: Harvard University Press.

Reiss, J. (2010). In favour of a Millian proposal to reform biomedical research. *Synthese, 177*(3), 427–447.

Resnik, D. B. (2021). Review of H. Radder: *From commodification to the common good: Reconstructing science, technology, and society*. Pittsburgh, PA: University of Pittsburgh Press, 2019. *Metascience*, 171–174.

Shah, S. (1993). *The body hunters: Testing new drugs on the World's Poorest Patients*. New York: New Press.

Shelley, C. (2006). Analogical reasoning with animal models in biomedical research. In L. Magnani (Ed.), *Model-based reasoning in science and engineering* (pp. 203–213). London: College Publications.

Shrader-Frechette, K. (2010). Conceptual analysis and special-interest science: Toxicology and the case of Edward Calabrese. *Synthese, 177*(3), 449–469.

Spinello, R. A. (1992). Ethics, pricing and the pharmaceutical industry. *Journal of Business Ethics 11*, 617–626.

Tuana, N. (2006). The speculum of ignorance: The women's health movement and epistemologies of ignorance. *Hypatia, 21*(3), 1–19.

Turner E., Matthews, A., Linardatos, E., Tell, R., & Rosenthal, R. (2008). Selective publication of antidepressant trials and its influence on apparent efficacy. *The New England Journal of Medicine, 358*, 252–260.

# Chapter 5
# The Future of Eco-Cognitive Settings

## Computationally or Humanly Tailored?

Taking advantage of the logical and cognitive studies illustrated in the previous chapters, which emphasize the crucial role played in abductive cognition by the so-called "optimization of eco-cognitive openness and situadedness", "knowledge in motion", and the concept of "epistemic irresponsibility", the present chapter will introduce the concept of overcomputationalism, to help interpret the related concepts of pancognitivism, paninformationalism, and pancomputationalism and their impact on discoverability. In the second part of the chapter I will submit to the attention of the reader a question that in my opinion synthesizes many of the problems described in this book: will the future of eco-cognitive settings computationally-tailored or humanly-tailored? The challenges against human abduction and epistemic rigor on the part of what I call computational invasive "subcultures" and unwelcome effects of selective ignorance are illustrated. Finally, I will describe two fundamental roles played by human knowledge and the need for their continuous enhancement sustaining discoverability: to the aim of maximizing the ownership of their own destinies, thanks to everyday, philosophical, and scientific knowledge about natural and artificial phenomena human beings have improved their capacities to track the external world; in turn thanks to ethical knowledge humans have learned to track the behavior of other human beings to the aim of favoring cooperation.

## 5.1 Protecting Ignorance: Beyond Overcomputationalization

### *5.1.1 Pancognitivism, Paninformationalism, Pancomputationalism*

What I call *overcomputationalization* refers to the presence of too many entities and artifacts that carry computational tasks and powers, for example, digital machines embedded in laptops, cell phones, servers, videos, cameras, social networks, robots,

 
L. Magnani, *Discoverability*, Logic, Argumentation & Reasoning 26,
https://doi.org/10.1007/978-3-030-93329-6_5

and also various unconventional substrates, etc.[1] Here the discourse becomes moral, and solutions are multiple. Whereas many would agree that the diffusion of computation is bringing havoc, some would say (like me) that it is caused by an excess of computation, others would content the opposite, i.e. computation has not spread enough.

Hence, our era is characterized by a progressive transformation of ignorant entities not only in cognitive entities but also in computational ones: I would like to suggest some humble considerations concerning the potential excesses of this trend, that I think have to be analyzed and criticized. Indeed, I think that currently we are facing an excess of computationalization of the world and I am consequently complaining that too many ignorant entities are more and more computationally "domesticated" (in the sense I have explained in this chapter). I will say more words on this issue, related to desire and suggestion of protecting the ignorance of some entities, in the following subsection of this chapter. There is no contradiction between this celebration of ignorance and my commitment to discoverability and diagnosticability because they are not favored by the excess of generic information.

In the last decades philosophers have celebrated the computational revolution and positively evaluated many of its aspects and consequences. An interesting position that has been elaborated is pancomputationalism that from a philosophical perspective, as we will soon see, is in some sense the rightful heir of paninformationalism and pancognitivism. To better grasp my concerns regarding overcomputationalization it is useful to analyze the intertwining between these positions.

Pancomputationalist philosophers have imagined that everything is computing. The problem of pancomputationalism can be nicely synthesized thanks to the following argumentation, provided by Horsman et al. (2014, p. 2). To answer the question *when does a physical system compute*? is very difficult and no accepted answer is currently available. There is not a formalism able to tell us whether a computation is occurring physically, and various confusions affect the discussion concerning non-standard forms of computation. It is obviously accepted that a laptop processing a Matlab calculation and a server running a LaTeX text are physical systems operating a computation, but "However, when we move beyond standard and mass-produced technology, the question becomes more difficult to answer. Is a protein performing a compaction computation as it folds? Does a photon (quantum) compute the shortest path through a leaf in photosynthesis? Is the human mind a computer? A dog catching a stick? A stone sitting on the floor?" If we say that all are computers we affirm that everything that possesses a physical status is executing a computation by virtue of its own existence. It is obvious to derive that from a conviction like this one, that sees a universe in which everything is a computer, the concept itself of physical computation becomes empty. Indeed "To state that every physical process is a computation is simply to redefine what is meant by a 'physical

[1] In (Magnani, 2021) I have described in detail the case of the so-called "morphological computation".

process'—there is, then, no non-trivial content to the assertion. A statement such as 'everything is computation' is either false, or it is trivial; either way, it is not useful in determining properties of physical systems in practice" (Horsman, Stepney, Wagner, & Kendon, 2014, p. 2).

Pancomputationalism is often defended explaining that when we state that everything is computational, we are simply using the word computational as a way of "interpreting" the system at stake, and of course everything can be interpreted using that metaphor. We have to say that this position creates confusion between computational modeling and computational explanation, as I will soon specify in the following paragraph. Other kinds of pancomputationalism, appropriately linked to paninformationalism, contend that the entire universe can be seen as computational and also arrive to maintain that information and/or computation possess a kind of priority with respect to physical materiality. Gordana Dodig Crnkovic (2011; 2013; 2017) proposes a richer info-computational and more circumscribed and integrated view as a synthesis of pancomputationalism—naturalist computationalism—with informational structural realism, and defends it by noting the fundamental role of computing in nature (natural computing).

As I have illustrated in Magnani (2018), adopting an evolutionary view and exploiting some of Thom's ideas expressed in the so-called catastrophe theory (Thom, 1988), paninformationalism is puzzling when seen in a naturalistic perspective. Let us consider the following example. When we are analyzing the case of an infection as a pregnance (carried by a virus, that is a material/biological mediator) that involves healthy agents, who represent the invested saliences that in turn can re-project that same contagion as a pregnance in a specific natural environment (where, in turn, other mediators such as air or blood are the transmitters), it seems bizarre to affirm that information (or even computation) is at play: biomedical scientific explanations of the phenomenon are already patently available and it seems superfluous to resort to other concepts such as the ones of information or of computation. One thing is to build a "model" of that event from an informational point of view or by using a computational program, and another is to produce a biological scientific knowledge of it. Of course, this critique does not jeopardize the project (again, see for example the info-computational view (Dodig-Crnkovic, 2017)) of enabling a view of nature also referred to on computational modeling (information processing) thanks to a conjugation of all those specialized approaches in a perspective that does not aim at replacing natural sciences with computational models, but that aims at establishing a fruitful epistemological collaboration.[2]

[2] Still in Magnani (2018), taking advantage of the studies provided by Longo (2009, 2012, 2017) I have described the dangers that can arise by thinking that mimetic computational digital modeling "is" directly, ipso facto, scientific knowledge: for example, in the case of biological organisms, the gap between simulation and intelligibility is very strong, also because the variability is dominant. A different view is advanced by Denning (2007): information processes and computation continue to be found abundantly in the deep structures of many fields. Computing is not—in fact, never was—a science only of the artificial, cf. also Denning & Martell (2015). A good example of the interplay between natural science and computational modeling is given by Fisher & Henzinger (2007),

However, defending paninformationalism and pancomputationalism can be good for two fundamental reasons. The first one relates to the centrality of natural human evolution, which presents that widespread activity of semiosis, I quoted above (including all kinds of signs, not only the propositional ones) that has been already generated by our ancestors at the times of the birth of the so-called material culture. Indeed humans have constructed those huge *cognitive niches* (Odling-Smee, Laland, & Feldman, 2003; Laland & Sterelny, 2006; Laland & Brown, 2006), characterized by the presence of informational, cognitive (and more recently, computational) processes (cf. chapter two, this book, Sect. 2.1). This evolutionary evidence explains the philosophers' attitude to grant some ontological status to information and computation. However, I contend that adopting an approach in terms of "disseminated computation",[3] helps consider pancomputationalism in a naturalized way, acknowledging its limits and tempering its ontological and metaphysical overstatements.

In this chapter there is no room for dealing with the intricate jungle of the various concepts of pancomputationalism: the reader will find a detailed and rich description of various kinds of pancomputationalism, unlimited, limited, causal, interpretivist, ontic, etc. in Piccinini (2017).[4] Here it is sufficient to note that Horsman's and his coauthors' hostility against pancomputationalism seems to refer to the "ontic" version of it, that is the idea that the universe itself is a computing system, and everything in it is a computing system too (or at least to the view that every physical system is a digital computing one and can be illustrated in computational terms).

We have to add that the strong advocates of the idea that universe computes contemplate a rather uncomplicated idea, and maintain that this happens thanks to different kinds of computation on different levels of organization, a contention that does not imply the acceptation that "everything is computation" (cf., for example, the already quoted info-computational view (Dodig-Crnkovic, 2017)), which would instead become a vacuous statement equivalent to the claim that nothing is computation. In this perspective, they say, by introducing the new notion of universal computation where "universe" is a whole universe and not only the symbol manipulation logical device, an epistemological gain is reached: for example, we can build new theoretical frameworks across disciplines that can help common understanding of the information processing in our brains and computations in the DNA or quantum computer. Grand unifications have always been attractive

who call the approach of constructing computational models of biological systems "executable biology", which focuses on the design of executable computer algorithms that mimic biological phenomena, an approach that the authors think must be integrated into biological research related to mathematical modeling.

[3] The expression is introduced and explained in section "The birth of mimetic bodies: Enhancing ignorant bodies through disseminated computation" of Magnani (2021).

[4] Pancomputationalism is also usefully discussed in the recent (Coelho Mollo, 2019), in the framework of an interesting and intricate comparison between the so-called "computational perspectivalism" and the mainstream accounts of physical computation, especially the teleologically-based mechanistic view.

in physics and mathematics and they are not an obstacle to appreciating diversity in which real world unfolds for us. Again, the strong advocates of the idea that universe computes, conclude that there would be absolutely no danger in broadening computational frameworks as much as it is justified by the behaviors of the systems that we study within the computational area of research.

After all, as contended by Kari & Rozenberg (2008) in the article "The many facets of natural computing", the most important property of computation consists in the multiplicity of its aspects and their efficacy. If a concept is capable of providing grand unification such as in the case of the concepts of matter, energy, information etc., it is worth to be used, developed, and explored. Finally, pancomputationalism is seen as a claim not critically dependent on the idea of discrete computation because the universe as an analog computer does not necessarily presuppose discrete computation. A main objection to this perspective is related to the following problem: one thing is modeling phenomena, systems, and data from a computational point of view, or exploiting new and unconventional "ignorant" substrates as computational tools, another is the adoption of the ontological idea that the objectivity itself has a computational nature (this last conviction leading to the so-called "ontic pancomputationalism", Piccinini (2017)). Analogously, as I have already said, one thing is to produce computational models for example of physical systems, another is to produce scientific knowledge of them.[5]

The second reason is cognitive and epistemological: in the literature of various disciplines often there is even an all-embracing concept of information, a kind of paninformationalism, in which physical (or biological) information is generalized to every state of a physical (or biological) system that is consequently defined as an information-pregnant state (Wolfram, 2002; Lloyd, 2006), which has to be considered relevant. Indeed, this approach has fecundated various excellent studies that physicists (for example Chiribella et al. (2012) and Goyal (2011)), and some logicians, have realized, for instance developing mathematical frameworks for considering quantum theory in the light of the principles of information processing. This information physics is an attempt to reconstruct physics based on information. These studies will not replace physics by computational theory of it, because those types of knowledge operate in different realms: physical computation and information physics would not replace but complement physics with new insights. These results are appreciable, I am just showing my worries about the possible abuse of the concepts of information (and of computation) in physics and biology, as I have already said above.

Finally, what about pancognitivism? An old-fashioned metaphysical theory, called synechism, contends that matter and mind are interrelated and in some sense it is not possible to differentiate them. The human cognitive mind would have germinated according to the same metaphysical laws that control the entire universe leading us to guess that the mind has an inclination (as Peirce contended) to find true abductive hypotheses regarding nature and the entire universe itself. As

[5] I have treated in detail this problem in Magnani (2018).

I have illustrated in Magnani (2009) it is Peirce himself that was convinced that "thought" is not only related to the brain: he said that it appears in the work of bees, crystals, and throughout the purely physical world. In sum, seeing information, and/or cognition, and/or computation everywhere presents some virtues but some limitations have to be established.

### *5.1.2 Protecting Ignorant Entities*

Since the birth of the so-called *material culture* (Mithen, 1996, 1999), when handaxes were made by Early Humans, the transformation of what I called ignorant entities in cognitive ones (and more recently in computational ones), also thanks to ancient techniques and modern technological advances, obviously had a tremendous positive impact on western civilization. However, as I have anticipated, a kind of warning has to be presented: I contend we are dealing in our era with what I call *overcomputationalization*. This means we are in the presence of too many entities and artifacts that carry computational tasks and powers. This implies some potential dangers: for example overcomputationalisation (1) furnishes too many opportunities to promote plenty of possible unresolvable disorganizational consequences in the activities of our societies and, (2) favors philosophical reflections that, taking advantage of the seduction exercised by old-fashioned metaphysical allures disguised as technical "analytical" studies depict an oversimplified vision of the world.

There is not enough space here to deal with the first of the two issues above but I can suggest that a lot of research in human sciences has been produced to deal with the ethical consequences of our technological computational era.[6] Instead, I would like to say few words concerning the second issue: we face a decrease in entities that remain simply "ignorant" from the informational/cognitive but especially from the computational point of view. Buildings, streets, furniture, clothes, etc. became carriers not only of simple traditional informational/cognitive obvious qualities but also, thanks to computationalization, of further complicated propositional and model-based information, sometimes related to redundant and oppressive political, ideological, and economical aims.

Moreover, more and more engineered and artefactual ignorant entities, laptops, cell phones, servers, videos, cameras, social networks, robots, and also various unconventional substrates, intentionally computationally domesticated are invading human environments. The recent relevance acquired by the studies on the so-called Internet of Things is a clear indication of the importance of these processes (Smart, Madaan, & Hall, 2019). The excesses of overcomputationalism can also generate a decrease of attention to entities when simply seen as more or less "ignorant" (that is devoid of complicated cognitive and computational endowments) and for this reason they tend to appear more and more irrelevant. The same happens in the

[6] See for example my own book (Magnani, 2007).

case of the excess of juxtaposition between new and complicated imposed informational/cognitive computational issues and various traditional entities. Unfortunately, these effects (1) tend to cause, in the case of overcomputationalism, too many constraints, limitations, and the weakening of discoverability and diagnosticability and so of human fruitful standard and creative (abductive) cognitive activities, as I have illustrated in this book and also in the last chapter of my book (Magnani, 2017) and, (2) in the case of the excess of redundant cognitive/informational features computationally attributed to entities (features in this case exogenous to the original functions of those entities) they incline to impede to benefit from the simplification that would derive thanks to a reduction of the consequent cognitive and informational overload, often potentially and negatively distracting.

The reader does not have to misunderstand me, I believe that we should choose carefully what to defend from computationalization (for example because it contrasts with the enhancing of discoverability in the case of human cognition), and what is good to computationalize, and so my urgency to "protect ignorant entities" does not have to be intended as a warning about all kinds of computationalization. Computational modeling of human body and brain can for example fosters a better understanding of diseases, degenerative processes, aging and other processes of life, including processes in our brains that generate the human mind. Human brain is the largely unknown organ in the human body and a lot of health problems, especially of aging populations is related to brains. Just to make few examples, we do not have cures for Alzheimer, Parkinson's disease, or Autism. So fruitful current and future computational modeling will not only favor further domestication of *ignorant* entities by transforming them in "skilled" substrates able to carry computational and so cognitive performances, but also will fecundate scientific research related to the removal of another kind of *ignorance*, the classical one regarding the lack of understanding of high-level cognitive capacities both in human, nonhuman animals, and simpler organisms, and other biological entities such as cells, as also hoped by the admonition contained in the recent review "Time to reinspect the foundations? Questioning if computer science is outgrowing its traditional foundations" (Copeland, Dresner, Proudfoot, & Shagrir, 2016).

## 5.2 The Future of Eco-Cognitive Settings: Computationally or Humanly Tailored?

Data and types of problems that are basically privileged by current mainstream AI programs are very restricted, data preferred are quantity-related (Big Data) and the best strategies that can be simulated are the *locked* ones, that limit the eco-cognitive openness and are typical of well-established games. It seems there is an increasing urgency to prepare data capable to feed the AI programs and to do that they have to be locked. This aspect generates a fundamental issue regarding ethics of algorithms and consequently, a final ethical warning has to be clearly stressed:

the widespread diffusion of standard and AI-based computational simulations will obviously incline human collectives to more and more adopt habits and settings that deal with data and problems appropriate to those same tools. This could generate an effect of progressive computational configuration of new social behaviors and environments appropriate to conform themselves to the computational tools and needs, to the detriment of old ones: in few words, will we face eco-cognitive settings which are *computationally tailored*, characterized by the prevalence of locked cognitive strategies or *humanly tailored*, characterized by the unlocked ones? I said that I call eco-cognitive openness the appropriate cognitive niche in which human creative cognition dwells, and this cognitive niche could be reduced and jeopardized by the computationalization of the world, so threatening the chance of getting new and important knowledge results (and not only in the case of science). Something similar could happen in the case of the prevalence of "locked"—in this case from the ethical point view—eco-cognitive settings in which the same chance of making human ethical decisions will be threatened, because objectively impeded by computational machines embedded in a triumphant "computationally tailored" eco-cognitive ethical situation.[7]

### *5.2.1 Computational Invasive "Subcultures" Undermine Human Creative Abduction in Science*

In the various sections of the previous chapter I have illustrated how various groups of technical, social, and political-economic forces can attract scientists' attention, constrain research and stimulate stakeholder interests along with particular directions. We have to say that also some computational research is directed by groups that aim at promoting the related business, unfortunately also challenging epistemic integrity, for example trivially contending their capacity to replace with artifacts human scientific creative abduction. To make a paradigmatic example Calude and Longo contend that

> Very large databases are a major opportunity for science and data analytics is a remarkable new field of investigation in computer science. The effectiveness of these tools is used to support a "philosophy" against the scientific method as developed throughout history. According to this view, computer-discovered correlations should replace scientific understanding as a guide to prediction and action. Consequently, there will be no need to give scientific meaning to phenomena, by proposing, say, causal relations, since regularities in very large databases are enough: "with enough data, the numbers speak for themselves". The "end of science" is proclaimed (Calude & Longo, 2017, p. 595).

[7] On the various "global catastrophic risk", from the actual and potential effects of climate change and nuclear war to plagues and pandemics, terrorism, biotechnologies, and nanotechnologies, but also on the potential dangers of artificial intelligence see the rich (Bostrom & Ćirković, 2008), in which the need of "making progress on friendly artificial intelligence" is claimed (on this issue see also Bostrom 2014).

The authors demonstrate, taking advantage of deep classical results from ergodic theory, Ramsey theory, and algorithmic information theory, how absurd is this contention and that instead very large database[8] present too many arbitrary—and so spurious—correlations, which surely cannot be considered examples of pregnant scientific creative abduction, but just uninteresting generalizations,[9] even if made thanks to sophisticated artifacts: "Too much information tends to behave like very little information" (Calude & Longo, 2017, p. 600). This study is the correct answer to the implicit challenge to scientific cognition proposed in June 2008 by C. Anderson, former editor-in-chief of *Wired Magazine*, who wrote an article titled "The end of theory: the data deluge makes the scientific method obsolete" contending that "with enough data, the numbers speak for themselves" (Calude & Longo, 2017, p. 595), science as we know will be replaced by robust correlations in immense databases![10]

Amazing examples of the use of these correlations are illustrated. The following is an eloquent example that indicates the extreme spontaneity and naturalness of some types of epistemic irresponsibility, especially in the case of the cognitive performances of rudimentary politicians and mass media journalists:

> A 2010 study conducted by Harvard economists Carmen Reinhart and Kenneth Rogoff reported a correlation between a country's economic growth and its debt-to-GDP ratio. In countries where public debt is over 90% of GDP, economic growth is slower, they found. The study gathered a lot of attention. Google Scholar listed 1218 citations at the time of writing, and the statistic was used in U.S. economic policy debates in 2011 and 2012 [...]. The study isn't conclusive, though-in fact, it's far from it. As noted by John Irons and Josh Bivens of the Economic Policy Institute, it's possible that the effect runs the other way round, with slow growth leading to high debt. Even more worryingly, the research didn't hold up on replication. But by the time that became clear, the original study had already attracted widespread attention and affected policy decisions (Calude & Longo, 2017, p. 599).

---

[8] Sax (2016) provides a discussion of the so-called libertarian-inspired "finders, keepers" ethics, that is the curious "morality" adopted by the business model of big data companies, which can "legitimately" appropriate (the fruits of) their results. Frizzo-Barker et al. (2016) also describe large-scale networked genetic material as a disruptive technology. On one hand, clinical genomics advances life-saving innovation through precision medicine. On the other, the digital databases that are built upon raise new concerns for informational risk to personal privacy.

[9] Such as the relation between the orientation of a comet's tail and the Emperor's chances of a military victory. Large database seem to rejoin magic, where underlying rational causes are lacking: but correlations do not supersede causation.

[10] More and more often researchers in different disciplines have been interested in using big data for their applications. Nonetheless, "For big data, spurious correlation refers to uncorrelated variables being falsely found to be correlated due to the massive size of the dataset" (Gandomi & Haider, 2015, p. 143). Studies that use big data are thus more likely to achieve the "generalizability" criterion of the findings, yet they are also prone to potentially produce uninteresting and/or insignificant, though generalizable, results due to the mentioned spurious correlations (see e.g. Fan et al. 2014).

The authors also note that European policy makers largely referred to that paper till 2013. The EU Commissioner for Economic Affairs—2009–13—referred to the Reinhart-Rogoff correlation as a key guideline for his past and present economic views and the British Chancellor of the Exchequer—since 2010—observed in April 2013 that Rogoff and Reinhart demonstrate "convincingly" that all financial crises have their ultimate origins in the public debt.

Let me remember another challenge to human abductive reasoning performed in the area of artificial intelligence (AI), very interesting but exaggeratedly emphasized by mass media, social networks, and the corporation itself, which proudly made it, that is Google. A long tradition in artificial intelligence—mainly in the area usually called "machine learning"—concerned the epistemologically very impressive computational AI applications that involve the abductive processes in scientific discovery and mathematical reasoning and creativity.[11]

In chapter three I have described the AI program that was able to play the Game Go very successfully, and so not involved in simulating scientific discovery but certainly human skillful strategic ability and creativity. It is well-known that in 2015 Google DeepMind's program AlphaGo beat Fan Hui, the European Go champion and a 2 dan (out of 9 dan possible) professional, five times out of five with no handicap on a full size $19 \times 19$ board. In March 2016, Google also challenged Lee Sedol, a 9 dan considered the top world player, to a five-game match. The program shot down Lee in four of the five games. I have clearly indicated the limitations of these kinds of computational systems.

Strong criticism of another invasive computational "subculture" is provided by Longo and Tendero (2007), who address their observations to the still widespread metaphor "DNA is a program" (the programming paradigm), used both in molecular biology and in its popularization. The authors contend that the metaphor and the model are fundamentally inadequate not only in biology but also from the point of view of both physics and computer science. Still in this case, analogously to the case of big data I have just illustrated, the programming paradigm is not theoretically sound as a causal framework for relating the genome to the phenotype, a much more complex process than the one which is depicted by an explanation based on this computational paradigm.

### 5.2.2 *Science Impoverished: Encouraging Epistemic Irresponsibility through Ignorance*

A general taxonomy of ignorance, occasioned by the feminist tradition in epistemology, is usefully provided by Tuana (2006). The various items below compose

[11] As already indicated in chapter three above I have discussed the problem of automatic scientific discovery with AI programs in Magnani (2009, chapter two, section 2.7 "Automatic Abductive Scientists").

the program of the so-called "epistemology of ignorance", which refers to a wide group of problems regarding human cognitive behavior, beyond the ones strictly referred to science. The list is interesting and some of the items can also regard science. However, I added three items that are more directly related the relationship between ignorance and scientific cognition:

1. "knowing that we do not know, but not caring to know", in this case ignorance is sometimes the result of the configurations of interest. I have already quoted the case of drug companies that decided research in the area of hormonal contraception for men would not be profitable (cf. Tuana (2006, p. 4)):
2. "we do not even know that we do not know" because our current interests, beliefs, and theories obscure some issues. A canonical example is provided by the feminists' notes about twentieth-century medical and anatomical illustrations which included careful examinations of male and female genitalia, but the anatomy of the clitoris was not a focus of attention and almost ignored in some mid-century textbooks;
3. "they do not want us to know", when the ignorance of certain groups is systematically cultivated. Some knowledge is rendered secret because of national security or because of business interests. I have to add that sometimes even people that know tend to ignore what they know. Indeed, an important question is not simply whether something is known but also who knows it and whether that knowledge is adequately exploited in the decision making techniques used by those in power: "Selective ignorance is important to emphasize even if it is not a matter of complete ignorance but rather a matter of the relative attention given to particular forms of information and research pathways" (Elliott, 2012, p. 335);
4. "wilful ignorance", which is related to the situation of people that do not know and do not want to know, such as in some forms of racism, when a strong ignorance of the lives and histories of those considered inferior is actively cultivated;
5. "ignorance produced by the construction of epistemically disadvantaged identities". In this case we see people that are constructed as untrustworthy or situations in which a deliberate maintenance of ignorance to guarantee privileges and power is activated: "What feminist epistemologists and science studies theorists have carefully demonstrated is that our theories of knowledge and knowledge practices are far from democratic, maintaining criteria of credibility that favor members of privileged groups" (Tuana, 2006, p. 13);
6. "loving ignorance", when accepting that it is better to ignore events, people aspects, situations, that can increase conflicts or produce violence or disfavor good projects;

   and I would also add three further items, which are of direct interest in the case of science:

7. "faking knowledge to preserve ignorance": I have already quoted the case in which some politicians and decision-makers address the solution of a problem toward scientific or technical fixes rather than social or political solutions. In this

case I have to say that scientific knowledge is not precisely faked, but, so to speak, just reverberated as possible (and needed). Thus, before scientists even begin to formulate their own specific questions about a research topic, a significant and value-laden decision may already have been made tacitly by citizens or policy makers-namely, the decision to treat the problem primarily as a set of questions for scientists and engineers to address. Of course there is a number of non-technical solutions to social problems, which instead cannot be solved thanks to science, even imagining for science the most shining future;

8. "epistemic irresponsibility": I have illustrated in the previous chapter many ways of jeopardizing science and its epistemological integrity, that is ways of promoting a scientific ignorance that would have been avoided adopting more epistemic rigor;
9. "distributed ignorance" through misinformation, widespread biased beliefs, ideological just-so stories pertaining to science and pseudoscience in science-related contexts, such as virtual online communities and—by now extremely degraded—mass-media. This threat to science (and scientific mentality) is illustrated in Arfini (2019): the diffusion of not verifiable scientific news through digital media can contribute to the formation of huge bubbles of shallow understanding in the laymen public.[12]

The recent (Kourany & Carrier, 2020) stressed various cases in which science itself can become a source of ignorance. Generation of ignorance is active when in presence of biased design and interpretation of experiments and empirical studies, such in the case of "false advertising" by climate change deniers; in the "virtuous" construction of ignorance—for example, by curtailing studies on race—and gender-related cognitive differences; ignorance as the unintentional result of choices made during the research process, especially when rules, motivations, and methodologies promote an insistence on the primacy of commercial effects.[13]

An amazing case of ignorance management regards a recent analysis of humor out of ignorance as a niche curation technique.[14] The Incongruity Theory of humor suggests that humor is triggered by the perception of ignorance, perceived as an incongruity. If we understand humor as a social modality of signaling an instance of ignorance which urges its resolution (in this case ignorance elicits amusement), the cognitive process can be seen as a way of enhancing the quality of the related cognitive niche by keeping the ignorant in line without excluding the others from participating to common niche-construction activities (Bertolotti, 2020).

---

[12] Dotson (2014) further analyzes "epistemic oppression" as a persistent epistemic exclusion (more or less intentionally driven) that hinders one's contribution to knowledge production: it can result from inefficient shared epistemic resources, like organizational schemata or instituted social imaginaries, which generate epistemic exclusions, from scarce or inadequate epistemic resources.

[13] On the identification and evaluation of the epistemic and social concerns raised by the private funding and performance of science and effect on ignorance studies see also Fernández Pinto (2015).

[14] On curation strategies cf. Sect. 2.3, chapter two, this book.

It is mandatory to make a reference to the role of ignorance as epistemic management that preserves the boundaries and relations of a rhetoric system, it is a way of dealing with information inconsistent with one's beliefs. In this case ignorance is also considered productive because it triggers new knowledge that works to make rhetoric systems more resilient to potential instability (Mays, 2021).

Finally, it is also important to note that there are cases in which people remain ignorant not because of a lack of information and not because of certain costs deriving from the problem of collecting knowledge, but because of the costs associated with knowledge itself (Williams, 2021). When the costs of acquiring knowledge surpass the benefits of possessing it, ignorance is rational. In this case ignorance is also "motivated": to make an example, it plays a large role in the high levels of voter ignorance in current democracies. The voter does not have any practical incentive to collect knowledge (for example on climate change) of the issues she votes on because her vote presents an insignificant impact on political decisions. This means that there are eco-cognitive settings that do not favor discoverability and diagnosticability and so potential new knowledge: instead they promote ignorance and, in this perspective, a rational one!

## 5.3 Enhancing Discoverability as the Condition of Possibility of Tracking the External World

We have just said at the end of the previous subsection that there are eco-cognitive settings that do not promote neither discoverability nor diagnosticability and so potential new knowledge because the costs of acquiring knowledge surpass the benefits of possessing it, and so ignorance is rational. This is of course a very rare and extreme case, because, as I have tried to illustrate in all the chapters of this book, favoring discoverability and diagnosticability, and consequently human creative and selective abductive capacities to generate new fresh knowledge, is a crucial target for human beings. As already explained in the preface, two objectives, and substantial reasons justify the need to preserve and enhance the conditions of possibility of both human creative abduction and abductive diagnostic virtues:

1. First of all, sustaining the eco-cognitive conditions of discoverability and diagnosticability is fundamental because only cognitive creativity can grant the implementation of the other kinds of ecology, which are urgently at play in our era so that the ecological improvement can in turn affects human creativity: this explains the fact I stressed the need an ecology of human creativity;
2. Second, fresh and innovative knowledge of various kinds is, and will be, necessary to furnish the appropriate cognitive tools for managing our technological era, full of artifacts of various types that are, in turn, mostly fruit of complicated scientific knowledge. It is more than clear that the maintenance and flourishing of technological societies requires a great deal of new scientific and ethical knowledge as well as modern approaches to moral deliberation.

To the aim of maximizing the ownership of their own destinies human beings have created a huge quantity of knowledge: first of all, thanks to everyday, philosophical, and scientific knowledge about natural and artificial phenomena human beings have improved their capacities to track the external world. In turn, thanks to ethical knowledge humans have learned to track the behavior of other human beings with the aim of favoring cooperation. I will describe in more detail these fundamental roles played by knowledge in the following subsections: as a consequence of their importance, in my book *Morality in a Technological World* (Magnani, 2007). I contended that knowledge has to be a duty. In the present book I have stressed the attention to a new meta-duty: to preserve, maintain, refine, and enhance these kinds of knowledge human beings also have to first of all take care of providing adequate room for discoverability and diagnosticability.

### *5.3.1 Tracking the External World I: Enhancing Predictive Knowledge*

I think that everyday, philosophical, and scientific knowledge about natural and artificial phenomena (as well as the technologies that relate to them) have allowed human beings a wide range of possibilities for choosing and acting: in many ways, the more one knows, the more options one has.

The evolution of knowledge and its externalization in objects and artifacts are directly related to our bodies' and brains' capacities for consciousness and free will and is directly related to the role of abduction. Indeed, knowledge sheds light on both natural and artificial external world, revealing two kinds of phenomena: (1) regular, foreseeable phenomena whose predictability makes possible a range of free choices and allows us to plan appropriate and effective responses and (2) unpredictable or unprecedented phenomena that cannot be altered by human intervention because we lack the knowledge to do so. Moreover, as we well know, this second sort of knowledge, because of its dynamics, does not involve "ontological" limitations: a phenomenon once considered beyond the reach of human action can, with new understanding, become easily manageable, a process that increases our options and, consequently, our free will. In the knowledge framework of some primitive people, for example, the course of a river was not considered modifiable, but new knowledge and new artifacts made it feasible to govern flowing water, allowing other ancient civilizations to "choose" between natural and artificial courses.

Dennett lists five different kinds of natural phenomena that relate to the "elbow room" he claims is the basic requirement for free will: those that are fixed, beneath notice, changing (and worth caring about),[15] trackable ("at least under some conditions—and hence efficiently and usefully predictable under those conditions"), and chaotic, that is (practically) unpredictable but still worth caring about). These categories of phenomena—which, I repeat, are dynamic products of human knowledge—provide us with what Dennett calls "epistemic possibilities", which are necessary for an agent to become a free deliberator. This variety encompasses everything that is "possible-for-all-the-deliberator-knows-or-cares" and ensures that every "deliberator-agent—a species, for instance—will always be equipped with a somewhat idiosyncratic way of gathering and partitioning information about its world so it can act effectively in it. [...] It is this epistemic openness, this possibility-for-all-one-knows, that provides the elbow room required for deliberation". This epistemic openness also nullifies deterministic objections to free will (Dennett, 1984, pp. 111 and 113). The tradition of western science maximizes the epistemic openness that provides the elbow room required for deliberation and so the chances of tracking the external world: in the previous chapters I have illustrated that science is characterized by a maximization of abducibility, performed thanks to specific constraints and through a process in which affirming truths implies negating truths.

Two simple examples will clarify the problem of the choice and its "elbow room" one related to everyday situations, the other related to science (cf. the following subsection). It is supposed I have a normal brain endowed with normal consciousness. Phylogenesis and (my personal) epigenesis have provided me with "hardware", the physical equipment that can be described as having free will because it can adopt various cognitive approaches to decision making, strategies human beings are now also able to model through algorithms (in my case, the non-technical ones I acquired and refined especially during my childhood).

At this moment, I am in my hotel room writing on my laptop; later today, I plan to go to the Sun Yat-sen University, which is relatively close. My everyday knowledge of the city (Guangzhou) (some of it is already stored in my memory) tells me that I can "choose" many routes to reach Sun Yat-sen University. I already have the mental representation of a possible route I used in the past, but I want to find a new one that is much shorter. All I need are new representations of the roads, which I can easily "pick up" from a map. Most of the world's information is stored in external "mediators"—it is not necessarily found only in brains!

At this point all the required elements for decision making are present: brain, consciousness, intentionality, free will, a reasoning mechanism guided by a system of values, and adequate knowledge about my goal and the possible routes I might take. Some aspects are internal and some are external. Some external ones—new data about possible routes, for example—become internal when they are put in memory and are therefore represented in my brain. I know there are routes I cannot choose because they involve loops or a *cul de sac*. At this point, I can plan my route.

[15] Dennett (1984, p. 109). Dennett also sees the chaotic systems as "the source of the 'practical' (but one might say infinitely practical) independence of things that shuffle the world and makes it a place of continual opportunity" (Dennett, 1984, p. 152).

At this stage, I can do more: I can externalize some data that are internal by drawing a simple map for the brain of a friend who needs information about how to get to the park. To an observer of this hypothetical interaction—a third brain—all aspects of this exchange between my friend and me are "external", and even my brain is considered an external object by the observer. But that third brain easily hypothesizes that my drawings derive from my existing "internal" representations. Cognitive psychologists contend that human consciousness requires a capacity to infer others' mental states from their behavior (Wegner, 2002). In turn, philosophers used those hypothesized mental states as evidence that being aware of our own internal representations derives from the immaterial Cartesian Cogito, that famous supplemental and embarrassing "ghost in the machine".

Let us come back to the problems of conscious free will and the role of knowledge, both internal and external. The simple example of my quest to reach Sun Yat-sen University has a clear epistemological significance; the new knowledge involved, both available and represented, indicates that based on the external mixture of natural things and artifacts that is my environment, "there are" various routes shorter than the one already stored in my mind. These shortcuts are then revealed and evaluated by the decision-making mechanisms operating in my brain. By picking up new knowledge from external devices (a map, for example) and re-representing it in my brain cells, I can enhance my range of possible choices and so gain more room to flex my free will.

As we can easily see, all the "characters" of this theater of consciousness are intertwined with one another (Seth & Baars, 2005). Consciousness in human brains (and so "my" consciousness) has evolved in this way because human brains, in turn, have produced knowledge about the world that sanctioned as available and believable multiple representations, that, once re-represented in the brain, further expand our menu of possible choices, and vice versa. Consequently, I contend that consciousness and higher mechanism of knowledge (such as abduction) are very much interdependent.

### *5.3.2 Tracking the External World II: The Role of Scientific Knowledge*

It is not only everyday knowledge that helps us to track the external world; higher levels of knowledge, like that found in philosophy and science, multiply the options available to human beings for choosing and acting. Consider for a moment a set of external natural circumstances and/or lack of knowledge that renders voluntary choice impossible—volcanic eruption, for example. Stored in my brain, which is endowed with consciousness function etc., is enough general knowledge of volcanology for me to understand that neither I nor any other human being can possibly predict the exact date of the next eruption of Vesuvius; even scientists who are experts in the field can only estimate that it will probably occur some time in the next 200 years. Consequently, in the next 200 years I cannot freely choose when to go to Naples and be certain I will avoid an eruption. Human beings currently have limited abilities to predict and thus to affect the impact of geological events in particular and of chaotic events in general—we cannot, therefore, reconfigure the Naples soccer team's schedule so a game does not coincide with the eruption. In this case, our capacity for free will is of no use: even with help from scientific models, computational devices, and complicated calculations, our brains are not able to pinpoint the precise day Vesuvius will erupt. Our best choice might be to postpone a trip to Naples until after 2222! Future improvements in volcanology, however, could allow more accurate predictions, even if we know that unpredictability is constitutive in chaotic phenomena.

If the world "out there" were always cognitively dark and homogeneous, free will would be impossible: free will and knowledge are two sides of the same coin. In recent human evolution, there has been a general increase in the human abductive production of philosophical and, subsequently, scientific knowledge (and in their systematic "externalization") as people have sought to free the human organism from its immediate environment. It is easy to imagine how many representations and inferential mechanisms can be stored in our brains and used at will: even those not already consigned to our brain memory exist over there, crystallized in various external mediators throughout the history of civilizations, ready to be "picked up" when needed. To that same end, human beings also participate in the reverse process by externalizing many techniques and technologies.

The roles of knowledge, however, extend beyond phenomena of the natural and artificial world, beyond settings like the streets of a city. In the external world there are also other human beings. What happens to our internal free will mechanisms when we are faced with the behavior of other human beings?

### *5.3.3 Tracking Human Behavior: Rendering Human Behavior Predictable Through Ethics*

In the previous two subsections I contended that cognitively tracking the external natural and artificial world provides the "elbow room" necessary to build a free deliberative agent, and that, unfortunately, one of the main obstacles to free choice (and thus to making free will effective) is not only the lack of suitable predictive knowledge about the processes of the natural and artificial world, but also about the behavior of other human beings. From this perspective, other people are "natural things" whose behavior is a priori difficult to predict: how can we track human intentions? Consequently, human behavior poses a very different sort of challenge, with respect to the constitutive incompleteness or lack of trustworthiness of available knowledge about natural and artificial phenomena. Indeed, when we "morally" seek ownership of our own destinies we expect to be able to reach objectives through consciousness, free will, and intentionality, thus undertaking a stand for the "best". We can only obtain the desired results if we can count on some consistency and predictability in the behavior of other human agents. If, in an attempt to "author" my destiny, I consider *merit* as a way to achieve the desired position, I must be able to assume that other human beings of my collectivity value it similarly. I contend that many objectified entities like common and religious morality, moral philosophy, human and social sciences, and of course all other kinds of ethical knowledge, are clearly connected to our existing need to operate at our highest level of conscious activity, as is the case when we seek to exercise free will and to claim ownership of our destinies.

How can I fruitfully employ my brain's free will mechanisms if I cannot trust other human beings? How can I work on a personal project or participate in a social project if not by relying on the commitment of other human agents? How may I "author" my life and reach my goals if I am unsure which actions to choose because I cannot be assured that others share my values and support my intentions? Religion, morality, moral knowledge, and teaching enhance and permit free will because they impose order on the randomness of human behaviors, giving people a better chance of owning their destinies.

There are many human actions that affect others' free will and ownership of destiny; among them, as recently observed by some scholars is the practice of gossiping. These authors affirm that the practice is not just an exchange of information about absent people, which can of course be a form of indirect aggression: it is also, they contend, a form of sociable interaction[16] processing others as "moral characters" (Yerkovich, 1977). The narratives created by gossiping become a possible source of shared knowledge about evaluative categories concerning (moral) ways of acting and interacting. Gossip does not have to be evaluative, but it is "moral" insofar as it describes behaviors and presents them as interesting and salient and, consequently, as potentially or *de facto* sanctionable. Gossiping could play an important role in morally—more or less violently—"policing free-riders" (Dunbar, 2004), that is, those who enjoy the benefits of sociality but refuse to pay their share of its costs. Commenting on the behavior of such people, or casting aspersions on their character helps us to control their potentially destructive effect on societies.

I have contended above that moral practices protect the ownership of our destinies because ethics render human behavior more predictable, and when we can count on shared values in dealing with other "moral" human beings, we can better project our future. Consequently, gossip helps safeguard the ownership of our destinies as it constantly shapes our narrative constructions of morality: empirical data have shown, for example, that gossip works as a form of low-cost (moral) social cognition that conveys valuable information about culture and society. The

[16] In my research I often emphasized the role of gossip. Dunbar (2004) originally gives scientific cognitive dignity to gossip explaining it in the framework of the so-called "social brain hypothesis". Posited in the late 1980s, this hypothesis contends that the relatively large brains of human beings and other primates reflect the cognitive demands of complex social systems and not just the need to process information of ecological relevance. Cf. also the recent (Magnani, 2011 and Bertolotti and Magnani, 2014).

act of gossiping can allow us to recognize that others are at risk of exploitation by moral free-riders even though we ourselves are not (Dunbar, 2004, pp. 106–109). Kant said that the "Kingdom of Ends"—that is, the moral world—"is a practical Idea used to bring into existence what does not exist but can be made actual by our conduct—and indeed to bring it into existence in conformity with this Idea" (Kant, 1964, p. 104).

Hence, the kingdom of ends is a kingdom of possible free choices created by (and contingent upon) human beings, for it is only their reliability that makes free will, and thus responsibility and freedom, possible. Dennett, when discussing the status of "self-made selves", makes the following comment: "Kant's famous claim in *Foundations of the Metaphysics of Morals* that the law we give ourselves does not bind us suggests that the selves we become in this process are not constrained by the law we promulgate because these selves are (partly) constituted by those very laws, partly created by a fiat that renders more articulate and definite something hitherto underdone or unformed" (Dennett, 1984, p. 90). Moreover, human aspects that are the underpinnings of the kingdom of ends—such as religion, for example—must be successfully and appropriately activated; their being in good working order is a basic condition for exercising morality and allowing free will to become "good" will. As I have stressed in my book *Morality on a Technological World* (Magnani, 2007), I am a sustainer of the importance of knowledge in constructing a new ethical commitment that embraces the idea of "respecting people as things",[17] and I always stressed the fact that particular kinds of technologies can threaten the growth—and even the existence—of freedom, responsibility, and the ownership of our destinies.

How could we use our free will without the constraints of objectified morality, religions, laws, and institutions that impose regularity and predictability on human behavior, and that, in turn, bolster people's trustworthiness? In this sense, we are responsible for our own free will (and, therefore, our freedom) because its existence and its perpetuity seem not to be an inalienable given, but they depend on our intellectual and practical choices about knowledge, religions, scientific and moral institutions and the related techniques and technologies, and on their use in everyday settings, work environments, education, communication, and economic life. For example, environmental imperatives are matters of principle that cannot be economically bargained away because they represent a kind of paradox of liberalism. Indeed, in matters of conservation, one could maintain that neutrality is necessary to preserve the rights of the individuals involved, but this notion is obviously outweighed by the fact that the freedom to destroy natural goods and things today will, paradoxically, inhibit freedom in the future, when people will have as a result fewer options when choosing among competing ideas of the good life.

[17] In essence, the idea holds that human beings often can and even should be treated as "things", and that in the process they become "respected as things" that had been ascribed more value than some people. We must reappropriate the instrumental and moral values that people have lavished on external things and objects, which I contend is central to reconfiguring human dignity in our technological world.

Similarly, as I have tried to demonstrate in this book, we are responsible for the preservation of high degrees of discoverability and diagnosticability, and therefore, for potential fruits of human abductive creative and selective cognition, because their abundance seems not to be an inalienable given too, but depends on our intellectual and practical curation of eco-cognitive openness.

## Summary

Enhancing discoverabiity and so human creativity is certainly related to the logical and cognitive studies I have addressed in the previous chapters, which highlighted many basic themes: "eco-cognitive openness and situadedness", "knowledge in motion", and "epistemic irresponsibility". The present chapter stressed the attention to the actual and possible negative consequences on diagnosticability and on discoverability (and so on human creativity) of that supposed excess of computationalization of the world that currently affects our societies. I have analyzed the concept of overcomputationalism also to the aim of providing a novel clarification of the concomitant concepts of pancognitivism, paninformationalism, and pancomputationalism. The last part of the chapter provided some possible insights that could be useful to answer the following question/warning: will the future of eco-cognitive settings computationally-tailored or humanly-tailored? Current threats against the flourishing of human creative abductive cognition and capable to possibly jeopardize epistemic rigor on the part of what I call computational invasive "subcultures" and unwelcome effects of selective ignorance have been illustrated. Finally, I have described two fundamental roles played by human knowledge: everyday, philosophical, and scientific knowledge about natural and artificial phenomena to the aim of tracking the external world; ethical knowledge to the aim of tracking the behavior of other human beings to favor cooperation.

## References

Arfini, S. (2019). *Ignorant cognition: A philosophical investigation of the cognitive features of not-knowing*. Cham: Springer.

Bertolotti, T. (2020). Humor and ignorance in the perspective of cognitive niche curation. In T. Bertolotti (Ed.), *Cognition in 3E: Emergent, embodied, extended: Multidisciplinary perspectives* (pp. 23–35). Cham: Springer.

Bertolotti, T., & Magnani, L. (2014). An epistemological analysis of gossip and gossip-based knowledge. *Synthese, 191*, 4037–4067.

Bostrom, N. (2014). *Superintelligence: Paths, dangers, strategies*. Oxford: Clarendon Press.

Bostrom, N., & Ćirković, M. M. (Eds.). (2008). *Global catastrophic risk*. Oxford: Oxford University Press.

Calude, C. S., & Longo, G. (2017). The deluge of spurious correlations in big data. *Foundations of Science, 22*(3), 595–612.

Chiribella, G., D'Ariano, G. M., & Perinotti, P. (2012). Quantum theory, namely the pure and reversible theory of information. *Entropy, 14*, 1877–1893.

Coelho Mollo, D. (2019). Against computational perspectivalism. *The British Journal for the Philosophy of Science, 72*(4), 1129–1153.

Copeland, J., Dresner, E., Proudfoot, D., & Shagrir, O. (2016). Time to reinspect the foundations? *Communications of the ACM, 59*(11), 34–38.

Dennett, D. (1984). *Elbow room: The variety of free will worth wanting*. Cambridge, MA: The MIT Press.

Denning, P. J. (2007). Computing is a natural science. *Communications of the ACM, 50*(7), 14–18.

Denning, P. J., & Martell, C. H. (2015). *Great principles of computing*. Cambridge, MA: The MIT Press.

Dodig-Crnkovic, G. (2011). Significance of models of computation, from Turing model to natural computation. *Minds and Machines, 21*(2), 301–322.

Dodig-Crnkovic, G. (2013). The info-computational nature of morphological computing. In V. C. Müller (Ed.), *Philosophy and theory of artificial intelligence* (pp. 59–68). Berlin, Heidelberg: Springer.

Dodig-Crnkovic, G. (2017). Computational dynamics of natural information morphology, discretely/continuous. *Philosophies, 2*(4), 23.

Dotson, K. (2014). Conceptualizing epistemic oppression. *Social Epistemology, 28*(2), 115–138.

Dunbar, R. (2004). Gossip in evolutionary perspective. *Review of General Psychology, 8*(2), 100–110.

Elliott, K. (2012). Selective ignorance and agricultural research. *Science, Technology, and Human Values, 38*(3), 328–350.

Fan, J., Han, F., & Liu, H. (2014). Challenges of big data analysis. *International Journal of Information Management, 1*(2), 293–314.

Fernández Pinto, M. (2015). Tensions in agnotology: Normativity in the studies of commercially driven ignorance. *Social Studies of Science, 45*(2), 294–315.

Fisher, J., & Henzinger, T. A. (2007). Executable cell biology. *Nature Biotechnology, 25*, 1239–1249.

Frizzo-Barker, J., Chow-White, P. A., Charters, A., & Ha, D. (2016). Genomic big data and privacy: Challenges and opportunities for precision medicine. *Computer Supported Cooperative Work, 25*(2–3), 115–136.

Gandomi, A., & Haider, M. (2015). Beyond the hype: Big data concepts, methods and analytics. *International Journal of Information Management, 35*(2), 137–144.

Goyal, P. (2011). Information physics—Towards a new conception of physical reality. *Information, 3*, 567–594.

Horsman, C., Stepney, S., Wagner, R. C., & Kendon, V. (2014). When does a physical system compute? *Proceedings of the Royal Society A, 470*, 1–25. 470 20140182. https://doi.org/10.1098/rspa.2014.0182. Published 9 July 2014.

Kant, I. (1964). *Groundwork of the metaphysics of morals* (1785) (3d Ed.). Reprint of 1956, edited and translated by H. J. Paton, Hutchinson and Co., Ltd., London (3d Ed.). New York: Harper and Row.

Kari, L., & Rozenberg, G. (2008). The many facets of natural computing. *Communications of the ACM, 51*(10), 72–83.

Kourany, J., & Carrier, M. (2020). *Science and the production of ignorance: When the quest for knowledge is Thwarted*. Cambridge, MA: The MIT Press.

Laland, K. N., & Brown, G. R. (2006). Niche construction, human behavior, and the adaptive-lag hypothesis. *Evolutionary Anthropology, 15*, 95–104.

Laland, K. N., & Sterelny, K. (2006). Perspective: Seven reasons (not) to neglect niche construction. *Evolution International Journal of Organic Evolution, 60*(9), 4757–4779.

Lloyd, S. (2006). *Programming the universe: A quantum computer scientist takes on the cosmos*. New York, NY: Knopf.

Longo, G. (2009). Critique of computational reason in the natural sciences. In E. Gelenbe & J.-P Kahane (Eds.). *Fundamental concepts in computer science*. London: Imperial College Press/World Scientific.

Longo, G. (2012). Incomputability in physics and biology. *Mathematical Structures in Computer Science, 22*(5), 880–900.

Longo, G. (2017). The biological consequences of the computational world: Mathematical reflections on cancer biology. arXiv.org, *eprint arXiv:1701.08085*.

Longo, G., & Tendero, P.-E. (2007). The differential method and the causal incompleteness of programming theory in molecular biology. *Foundations of Science, 12*(4), 337–366.

Magnani, L. (2007). *Morality in a technological world: Knowledge as duty*. Cambridge: Cambridge University Press.

Magnani, L. (2009). *Abductive cognition: The epistemological and eco-cognitive dimensions of hypothetical reasoning*. Heidelberg/Berlin: Springer.

Magnani, L. (2011). *Understanding violence. The intertwining of morality, religion, and violence: A philosophical stance*. Heidelberg/Berlin: Springer.

Magnani, L. (2017). *The abductive structure of scientific creativity: An essay on the ecology of cognition*. Cham, Switzerland: Springer.

Magnani, L. (2018). Computationalism in a dynamic and distributed eco-cognitive perspective. In G. Dodig-Crnkovic & M. Burgin (Eds.), *Philosophy and methodology of information* (Vol. 1, pp. 365–288). Singapore: World Scientific.

Magnani, L. (2021). Computational domestication of ignorant entities. Unconventional cognitive embodiments. *Synthese, 198*, 7503–7532. Special Issue on "Knowing the Unknown" (guest editors L. Magnani and S. Arfini).

Mays, C. (2021). Ignorance as a productive response to epistemic perturbations. *Synthese, 198*, 6491–6507.

Mithen, S. (1996). *The prehistory of the mind: A search for the origins of art, religion, and science*. London: Thames and Hudson.

Mithen, S. (1999). Handaxes and ice age carvings: Hard evidence for the evolution of consciousness. In A. R. Hameroff, A. W. Kaszniak, & D. J. Chalmers (Eds.), *Toward a science of consciousness III: The third tucson discussions and debates* (pp. 281–296). Cambridge: MIT Press.

Odling-Smee, F. J., Laland, K. N., & Feldman, M. W. (2003). *Niche construction: The neglected process in evolution*. Princeton, NJ: Princeton University Press.

Piccinini, G. (2017). Computation in physical systems. In E. N. Zalta (Ed.), *The stanford encyclopedia of philosophy* (Summer 2017). Stanford: Metaphysics Research Lab, Stanford University.

Sax, M. (2016). Big data: Finders keepers, losers weepers? *Ethics and Information Technology, 18*(1), 25–31.

Seth, A. K., & Baars, B. J. (2005). Neural Darwinism and consciousness. *Consciousness and Cognition, 14*, 140–168.

Smart, P., Madaan, A., & Hall, W. (2019). Where the smart things are: Social machines and the internet of things. *Phenomenology and the Cognitive Sciences, 18*(3), 551–575.

Thom, R. (1988). *Esquisse d'une sémiophysique*. Translated by V. Meyer, *Semio Physics: A Sketch*, Addison Wesley, Redwood City, CA, 1990. Paris: InterEditions.

Tuana, N. (2006). The speculum of ignorance: The women's health movement and epistemologies of ignorance. *Hypatia, 21*(3), 1–19.

Wegner, D. M. (2002). *The illusion of conscious will*. Cambridge, MA: MIT Press.

Williams, D. (2021). Motivated ignorance, rationality, and democratic politics. *Synthese* (Forthcoming), 1–21.

Wolfram, S. (2002). *A new kind of science*. Champaign: Wolfram Media.

Yerkovich, S. (1977). Gossip as a way of speaking. *Journal of Communication, 27*, 192–196.

# Conclusion

In place of a formal conclusion, I offer here a sort of final comment about what I see as the most important elements of *Discoverability*. I hope that I was able, through my arguments, to encourage a serious commitment to increase our knowledge about the importance of seeing the whole eco-cognitive aspects of either current incentives or threats to discoverability, as a philosophical, epistemological, cognitive, and ethical topic. The analysis of abductive cognition (especially creative) that I tried to operate focused on the analysis of its "conditions of possibility", *as they are* and *as they should become* to the aim of enhancing their efficiency and power. I abstained from explaining them *away* by any reduction to a rigid unilateral perspective, instead, I favored a multifarious analysis to improve our chances to maintain the theoretical commitment towards the study of human cognitive creativity as a fundamental endowment, profoundly intertwined with human beings' well-being and their constructive chances of maintaining and increasing their capacity to effectively govern the strictly concomitant ecological problems. I hope I have been able to furnish an examination of discoverability in science because I think that having a better knowledge of it not only is a task endowed with an intellectual and didactic value, but also with "social" virtues: a fresh knowledge that promotes scientific human creativity at the same time is the condition of possibility of increasing the growth of other kinds of knowledge, which is desperately needed in our technological era.

At the same time, given the importance of creativity as a crucial instrument necessary to provide novel knowledge able to fix the failures of our societies and our lives, studies on discoverability help to maintain an intellectual focus on "moral" and "ecological" commitments, hopefully, also beyond the restricted intellectual community. The knowledge about discoverability I fervidly treated in this book also aims at supplying researchers with the philosophical stance required to manage disputable issues regarding the intertwining between freedom of research, eco-cognitive openness of epistemic environments, knowledge in motion, affordances, epistemic responsibility and irresponsibility, overcomputationalism, etc. and what



L. Magnani, *Discoverability*, Logic, Argumentation & Reasoning 26,
https://doi.org/10.1007/978-3-030-93329-6

I called "optimization of eco-cognitive openness and situadedness", for example jeopardized by the overwhelming commercialization of the results of various fields of scientific research.

More or less hypothetical abductive creative thinking is so pervasive, so much a part of human life, that it is not easy to imagine how to live without being positively affected by its benefits. Charles Sanders Peirce magnificently said that "Our faculty of guessing corresponds to a bird's musical and aeronautic powers; that is, it is to us, as those are to them, the loftiest of our merely instinctive powers". The huge impact that the constructive fruits of the human capacity to generate good hypotheses bear (and can further bear) on civilizations, but also on animals and nature in general, surely suggests that enhancing discoverability is the fundamental condition of possibility of creativity and at the same time the first ecological duty. Sustaining the eco-cognitive conditions of discoverability is fundamental because only creativity can grant the implementation of the other kinds of ecology, which are urgently at play in our era so that the ecological improvement can in turn affects human creativity. As I already said in the preface I consider the protection of discoverability as the tool for building an *ecology of human creativity* as a *conditio sine qua non* in the following sense: I am confident that without building appropriate high degrees of discoverability (to favor creativity and skillful human capacities) all the other envisaged and invoked ecologies and sustainability, in general, tend to miserably fail.

People can make use of different hypothetical abductive cognition as weapons of better well-being, but also to harm others—both for defensive and aggressive scopes—as I for example described along with the various chapters of my previous book *Understanding Violence*. However, poor creative cognitive capacities and results are certainly harbingers of negative effects: our technological era is full of artifacts of various type, all fruits of the "objectification" of scientific knowledge, and to manage them fresh and new knowledge in general (and scientific knowledge in particular) is always fundamental to maintain their functions and to protect humans from adverse and unintended negative damages. I tried to demonstrate the importance of individuating cognitive firewalls able to prevent the decline of discoverability, to further weaken this damage and to improve our human abductive creativity, freedom, and safer ownership of our destinies.

In this book I wanted to provide an analysis, in the light of my eco-cognitive approach to epistemology, of various threats to science—that I consider extremely negative—to human creative abduction, and to the related ways of producing rational knowledge developed in the last few centuries. Discoverability and diagnostcability seem jeopardized: consequently these threats seem to design a potential future of decivilization, or at least a future characterized by different forms of "civilization", incommensurable with the one of democracies we have experienced in the last part of the previous century in the western world. Unfortunately, it is highly difficult to address in a socially responsible manner the problems originated by these transformations. To recognize some of these threats is already making a good intellectual job, even if I am sure that many of these menaces, which are already active, are still difficult to be detected or completely out of intellectual sight.

In sum, understanding discoverability in an eco-cognitive perspective can be of help, for example, to avoid falling into a situation when humans become passive receivers of effects generated by technologies and/or ecological impairments in an environment marked by a poorness of solutions able to manage pressing and hazardous modifications, with consequent damage to the ownership of their destiny. In sum, seeing discoverability as an essential eco-cognitive factor can help to grasp the role of human creativity as a quality that cannot be disregarded and that is worth be examined not only in the perspective of ethical frameworks regarding the respect of human intelligence, merit, and high-level cognitive capacities but also as an objective epistemological and ecological issue, which regards the well-being of the humans as intertwined with nature.

# Lexicon of Discoverability

**Abductive Logic Programming** Abductive logic programming (ALP) is a high level knowledge-representation framework that can be used to solve problems declaratively based on abductive reasoning. It extends normal logic programming by allowing some predicates to be incompletely defined, declared as abducible predicates. Problem solving is accomplished by deriving hypotheses on these abducible predicates (abductive hypotheses) as solutions of problems to be solved. These problems can be either observations that need to be explained (as in classical abduction) or goals to be achieved (as in normal logic programming). It can be used to resolve problems in diagnosis, planning, natural language and machine learning. It has also been used to handle negation as failure as a way of expressing some forms of abductive reasoning.

**Adversarial Abduction** Adversarial abduction refers to deception as the activity of intentionally misleading other agents' hypothetical inferences. Deception is in this case analyzed as an intentional manipulation of others' hypothetical inferences so that they will accept the false or disadvantageous abduced hypothesis. This topic is also related to the role of the so-called adversarial cognition in intelligence service, in which the main aim is to abductively detect and exploit adversarial affordances, as environmentally situated opportunities for knowledge and action defined by a mutual relationship between adversaries. In these cases it is diagnosticability more than discoverability at play, and various ways of jeopardizing selective abduction (diagnosis) of other human beings are active, so as they consider they abductive performances reliable and good, but they are bad results in the third-person

In this "Lexicon of Discoverability" also certain aspects of abductive cognition that are left implicit in this book are illustrated. Further aspects of abduction are included in the more extended "Lexicon of Abductive Cognition" of my earlier book *Abductive Cognition. The Epistemological and Eco-Cognitive Dimensions of Hypothetical Reasoning* (2009).

 
L. Magnani, *Discoverability*, Logic, Argumentation & Reasoning 26,
https://doi.org/10.1007/978-3-030-93329-6

perspective of the deceiver that collaborated to (usually intentionally) generate them to get some advantages.

**Affordances and Abducibility/Discoverability/Diagnosticability** An affordance is a resource or chance that the environment presents to the specific organism, such as the availability of water or of finding recovery and concealment. From an abductive perspective, affordances as environmental chances can be related to the variable (degree of) abducibility (discoverability, diagnosticability) of a semiotic configuration of "signs": a chair affords sitting in the sense that the action of sitting is a result of a sign activity in which we perceive some physical properties (flatness, rigidity, etc.), and therefore we can ordinarily infer (in Peircean sense)—abduce—that a possible way to cope with a chair is sitting on it.

**AKM-Schema** The classical schematic representation of abduction is expressed by what (Gabbay & Woods, 2005) call AKM-schema, which is contrasted to their own (GW-schema). For *A* they refer to Aliseda (1997, 2006), for *K* to Kowalski (1979), Kuipers (1999), and Kakas et al. (1993), for *M* to Magnani (2001) and Meheus et al. (2002). The AKM-schema illustrates the explanatory dimension of abduction. That is, it describes abduction as the kind of reasoning which aims at generating hypotheses, which are related to considerations of plausibility, relevance, and characteristicness. In this perspective abduction is immediately considered a generation/selection of "plausible" hypotheses: minimality and consistency of the results are mandatory. However, the AKM model does not exhaust the concept of abduction, which also presents a non-explanatory, instrumental, and manipulative dimension, possibly considered as not intrinsically consequentialist. AKM-schema can be illustrated as follows:

1. $E$
2. $K \not\rightsquigarrow E$
3. $H \not\rightsquigarrow E$
4. $K(H)$ is consistent
5. $K(H)$ is minimal
6. $K(H) \rightsquigarrow E$
7. Therefore, $H$
   (Gabbay & Woods, 2005, pp. 48–49)

where of course the conclusion operator $\rightsquigarrow$ cannot be classically interpreted.[1]

[1] The target has to be an explanation and $K(H)$ bears $R^{pres}$ [that is the relation of presumptive attainment] to $T$ only if there is a proposition $V$ and a consequence relation $\rightsquigarrow$ such that $K(H) \rightsquigarrow V$, where $V$ represents a *payoff proposition* for $T$. In turn, in this schema explanations are interpreted in consequentialist terms. If $E$ is an explanans and $E'$ an explanandum the first explains the second only if (some authors further contend if and only if) the first implies the second. It is obvious to add that the AKM schema embeds a D-N (deductive-nomological) interpretation of explanation, as I have already stressed in (Magnani, 2001, p. 39).

**Anticipation as Abduction** In the tradition of phenomenology anticipations correspond to a kind of an intentional expectation. When we see a spherical form from one perspective, we will assume that it is effectively a sphere, but it could be also a hemisphere. Anticipations share with other kinds of abductions various features: they are highly conjectural and nonmonotonic, so wrong anticipations have to be replaced by other plausible ones. Moreover, they constitute an activity of "generating and testing" as a kind of action-based cognition. Not all the anticipations are informationally equivalent and work like attractors for privileged individuations of objects. In this sense the whole activity is toward "the best anticipation", the one that can display the object in an optimal way.

**Anomalies as Triggers to Discovery** Anomaly refers to unexpected and uncharacteristic events that create problems which demand a solution. Anomalous events are marked by both an epistemic disadvantage and an emotional rating. Philosophers of science in the last century have illustrated that inconsistencies and anomalies often play an important role in the growth of scientific knowledge. Hence, contradictions and inconsistencies are fundamental in abductive reasoning, and abductive reasoning is appropriate for "governing" inconsistencies.

**Aristotle's ἀπαγωγή** Aristotle clearly states that in syllogistic theory local/environmental cognitive factors—external to that peculiar inferential process, for example regarding users/reasoners—are given up. At the same time in chapter B25 of the *Prior Analytics* Aristotle presents a seminal perspective on abduction: some of the current well-known distinctive characters of abductive cognition, and of abductive cognition in science, are already expressed, which are in tune with the Eco-Cognitive Model (EC-Model). By referring to the role of the method of analysis and of the middle terms in Plato's dialectic argumentation, considered as related to the diorismic/poristic process in ancient geometry, Aristotle is still pointing to the fundamental inferential and "distributed" role in reasoning of those externalities that substantiate the process of "leading away" (an expression which also translates what Aristotle calls ἀπαγωγή, that is "abduction").

**Autoimmunization** When inferential strategies are *locked*, both in human and artificial cognition in a determined perspective where the components do not change (for example in the case of games such as Go or Chess) an *autoimmunization* is active, that constitutes the limitations that preclude the application of strategies that are not related to "pre-packaged" scenarios, strategies that would be foreigners to the ones that are strictly intertwined with the components of the given scenario.

**Chance, Discoverability, Diagnosticability** Humans can be considered chance seekers as far as they are continuously engaged in a process of building up and then extracting latent possibilities to configure discoverability and diagnosticability and so to possibly uncover new valuable information and knowledge to trigger abductions. That is, humans like other organisms do not simply live their environment, but they actively shape and change it looking for suitable chances. Hence, in this perspective chances constitute that "information" which is not stored internally in memory or already available in an external reserve, but that has to be "extracted" and then picked up upon occasion.

**Confinement** Cognitive confinement refers to a contingent, yet relatively stable state of being structurally or systematically unable to gain information from an environment, determined by patterns of interaction between the subject and the world. It is considered as a pathologized form of cognitive niche: confinement affects human and non human animal eco-cognitive situatedness and so the abductive potential. If related to the concept of affordance, cognitive confinement is a state of being systematically blocked from affordances or even disafforded, as it were, and so provided with disfigured affordances due to the way in which the subject's environment (cognitive niche) is set up by the subject-world interactions.

**Curation of Eco-Cognitive Situatedness** Curation of eco-cognitive situatedness refers to all the activities capable to preserve, maintain, and enhance those aspects that favor discoverability, for example in science. Indeed, to favor scientific discovery, it has to be stressed that the optimization/maximization of eco-cognitive situatedness derives from respecting various constraints, methods, cognitive virtues, etc., which certainly substantiate limitations of the related cognitive behaviors but at the same time they are clearly finalized to the maintenance of the optimization/maximization of eco-cognitive openness, that solely can grant the flourishing of abductive scientific creativity. In other cases curing the eco-cognitive situatedness is related to the fact that chances as conditions of diagnosticability and discoverability can be faked. It seems intuitive to think that the suppression of chance-faking contexts is an important activity of chance curation.

**Creative Abduction** Creative abduction is the process in which a completely new hypothesis is created. An example of creative abduction is scientific discovery: the discovery of a new disease and the manifestations it causes in the field of medical knowledge or the Kepler's discovery of the elliptic orbit of the planets.

**Deduction and Abduction** Deduction is that kind of truth-preserving reasoning which starts from reasons and looks for consequences. In doing so deduction does not produce any new idea. New ideas are due to abduction that standardly involves the generation and evaluation of hypotheses. In the process of hypothesis generation and evaluation, deduction plays a fundamental role in deriving the consequences of hypotheses, which, in turn, are compared with the available data by induction.

**Diagnosis** Conjectures can be either the fruit of an abductive selection in a set of pre-stored hypotheses or the creation of new ones. "Creative" abduction deals for example with the whole field of the growth of scientific knowledge. This is irrelevant in medical diagnosis where instead the task is to "select" from an encyclopedia of pre-stored diagnostic entities. In this case abduction selects hypotheses; from these hypotheses consequences are derived by deduction that are compared with the available data by what Peirce called induction, in this case considered a method of evaluation.[2] Often the information available does not allow a physician to make a

[2] Peirce was right in denying the role of "naked" induction in forming new hypotheses and considered the term to indicate processes of evaluation. However, Hintikka usefully warns us about

precise diagnosis. Therefore, she has to get further data, or even try some different manipulations to uncover symptoms otherwise hidden, so favoring diagnosticability.

**Disfigured Affordances** Disfigured affordances are affordances that are blocked or, so to speak, disafforded, due to the way in which the subject's environment (cognitive niche) is set up by the subject-world interactions. The perception of disfigured affordances favors a possible course of action, e.g., the actions that someone might associate with bent sticks, whereas in fact the actions in question are not a real option.

**EC-Model and Knowledge Enhancing Abduction** The backbone of the approach to abduction illustrated in this book can be found in the original manifesto of my Eco-Cognitive Model (EC-Model) of abduction in (Magnani, 2009). This model always emphasizes the cognitive nature of abduction. Reasoning is something performed by cognitive systems. At a certain level of abstraction and as a first approximation, a cognitive system is a triple $(A, T, R)$, in which $A$ is an *agent*, $T$ is a *cognitive target* of the agent, and $R$ relates to the *cognitive resources* on which the agent can count in the course of trying to meet the target-information, time and computational capacity, to name the three most important. The agents are also considered as *embodied distributed cognitive systems*: cognition is embodied and the interactions between brains, bodies, and external environment are its central aspects. Cognition is occurring taking advantage of a constant exchange of information in a complex distributed system that crosses the boundary between humans, artifacts, and the surrounding environment, where also instinctual and unconscious abilities play an important role. This interplay is especially manifest and clear in various aspects of abductive cognition. The EC-Model of abduction provides the opportunity to illustrate the knowledge enhancing character of abduction, beyond its ignorance preserving character, stated by the GW-schema: through abduction, knowledge can be enhanced, even when abduction is not considered an inference to the best explanation in the classical sense of the expression, that is an inference necessarily characterized by an empirical evaluation phase, or an inductive phase, as Peirce called it.[3]

**Eco-Cognitive Openness, Discoverability, Abduction** A new formal perspective on abductive processes, in terms of the effort to naturalize the logic of its special consequence relation, leads to the emphasis on the importance of the following main

---

the use of the word induction in the case of the testing of hypotheses: "[...] I do not think that it is instructive to call such reasoning inductive, but this is a merely terminological matter" (Hintikka, 2007, p. 55).

[3] Peirce provides various justifications of the knowledge enhancing role of abduction: these justifications basically resort to the conceptual exploitation of *evolutionary* and *metaphysical* ideas, which clearly show that abduction is constitutively akin to truth, even if certainly always ignorance-preserving or mitigating in the sense that the "absolute truth" is never reached through abduction. Examples that indicate the knowledge enhancing role of abduction are: (1) abducing conventions in empirical science and (2) the process of abducing those models in scientific reasoning which play the "constitutive" function of enhancing knowledge.

aspects: "optimization of eco-cognitive situatedness", "maximization of changeability" of both input and output of the general form of an inferential abductive problem, and high "information-sensitiveness". Furthermore, a naturalized logic of abduction must acknowledge the importance of keeping record of the "past life" of abductive inferential praxes, contrarily to the fact that traditional demonstrative ideal systems are prototypically characterized by what I call "maximization of memorylessness". In this perspective it is important to stress what can be called *maximization of abducibility*, which is typical of scientific discovery. Constraints, methods, cognitive virtues, etc. which characterize scientific cognition certainly substantiate limitations of the cognitive behaviors but at the same time they are clearly finalized to the maintenance of the optimization/maximization of eco-cognitive openness, that solely can grant the flourishing of abductive scientific creativity. Situatedness is related to eco-cognitive aspects: to favor the solution of the abductive problem input and output of the formula

$$\Lambda_1, \ldots, \Lambda_i, ?_I \Vdash_L^X \Upsilon_1, \ldots, .\Upsilon_j$$

have to be thought as *optimally positioned*.[4] This optimality is in turn made possible in science by a *maximization of changeability* of both input and output. Not only inputs have to be enriched with the possible solution but, to do that, other inputs have usually to be changed and/or modified. Consequently, abductive inferential processes are highly *information-sensitive*, that is the flux of information which interferes with them is continuous and systematically human(or machine)-promoted and enhanced when needed. This is not true of traditional inferential settings, for example proofs in classical logic, in which the modifications of the inputs are *minimized*, proofs are usually taken with "given" inputs, and the burden of proofs is dominant and charged on rules of inferences, and on the smart choice of them together with the choice of their appropriate sequentiality; indeed, in the eco-cognitive perspective, an "inferential problem" can be enriched by the appearance of new outputs to be accounted for and the inferential process has to restart. This is exactly the case of abduction and the cycle of reasoning reflects the well-known nonmonotonic character of abductive reasoning. Abductive consequence is ruptured by new and newly disclosed information, and so defeasible. In this perspective abductive inference is *not only* the result of the modification of the inputs, but, in general, actually involves the intertwined modification of both inputs and outputs.

**Ecology of Human Creativity** If creativity is jeopardized and in particular scientific creativity, it is practically impossible to acceptably and successfully deal with our technological societies, also from an ethical perspective: this means that the need of an *ecology of human creativity* is mandatory.

---

[4] $\Vdash_L^{\surd}$ will denote that the expected relation is obtained. This formula is illustrated in chapter one, p. 16.

**Embodied and Distributed Cognition** Extra-theoretical aspects and manipulations of "external" objects in reasoning are very important in hypothetical reasoning. Paying attention to the perceptual and manipulative dimension of cognition, Peirce already reminded us that human cognitive systems are not "isolated" and autonomous entities. Cognition is embodied and the interactions between brains, bodies, and external environment are its central aspects. Cognition is occurring taking advantage of a constant exchange of information in a complex distributed system that crosses the boundary between humans, artifacts, and the surrounding environment. This interplay is especially manifest and clear in various aspects of abductive cognition.

**Emotion, Abducibility, Diagnosticability** Emotion furnishes immediate abductive appraisals of the bodily states, and provides a kind of explanation of them. Happiness, sadness, fear, anger, disgust, and surprise all can be viewed as judgments about a person's general state; a man who unexpectedly comes across a tiger on the loose, for example, would be understandably afraid because the large carnivore threatens his instinct to stay alive. In this sense, all emotions are connected to goal accomplishment: people become angry when they are thwarted, for instance, and feel pleased when they are successful.

**Epistemic Responsibility/Irresponsibility and Creative Abduction** The examination of some important aspects of the current organization of Research and Development (R&D), for example in the case of biopharmaceutical companies, represents a prototypical case of creation of the so-called impoverished epistemic niches. In these cases I see a challenge to the epistemic integrity and rigor of modern science and to the productivity of scientific abductive creativity. Currently a hot problem is the emergence of various kinds of "*epistemic irresponsibility*", such as in the case of the increasing commodification and commercialization of science, the marketing of technoscientific products, the emergence of computational invasive "subcultures", and the unwelcome effects of selective ignorance. The subsequent impoverishment of the so-called epistemological niches shows that discoverability is jeopardized and consequently human fruitful abductive cognition is increasingly assaulted, and at the same time human creativity seriously endangered.

**Fake Discoverability and Fake Diagnosticability** Chances as conditions of diagnosticability and discoverability can be faked. It seems intuitive to think that the suppression of chance-faking contexts is an important activity of chance curation. We can consider the amazing example of bullshit as a source of fake chances, but at the same time also—rarely—as a fertile ground for learning and developing some creative intuitions: a rich cognitive niche may benefit from certain kind of fake chances—which should therefore not be inhibited—whereas a poorer niche might not benefit from these fake aspects, and therefore the preclusion of fake chances is an important act of chance curation to protect discoverability and dignosticability.

**Fallacies and Abduction** In the perspective of classical logic abductive reasoning can be defined as fallacious. More precisely, abduction is classified as the fallacy of affirming the consequent. In the light of an agent-based approach to logic the fallacious character of abduction can be further clarified and weakened: abduction is recognized as a very precious method of explanation and discovery in science and in everyday reasoning. Hence, in an agent-based perspective this kind of fallacious reasoning can be redefined and considered as a fundamental and good way of reasoning.

**Fallacy of Affirming the Consequent** Abduction appears to be a formal fallacy that can be recognized in the perspective of classical logic: it is the fallacy of affirming the consequent. Basically it is an invalid inference, which assumes the form: "if $A$, then $B$, $B$; then $A$. From the classical logic point of view this inference is fallacious and not truth preserving.

**Filter Bubbles** The following example of the so-called "filter bubbles" is of help: access to online information—in the so-called virtual cognitive niches—is constrained by algorithms that filter content in accordance with the subject's previously detected preferences or biases, and this of course diminishes the likelihood of getting new or unexpected information. In this sense filter bubbles are special cases of a more general phenomenon of cognitive confinement related to processes of "ignorance-making".

**Future of Eco-Cognitive Settings** Data and types of problems that are basically privileged by current mainstream AI programs are very restricted, data preferred are quantity-related (big data) and the best strategies that can be simulated are the *locked* ones typical of well-established games. It seems there is an increasing urgency to prepare data capable to feed the AI programs and to do that they have to be locked. This aspect generates a fundamental issue regarding ethics of algorithms and consequently, a final ethical warning has to be clearly stressed: the widespread diffusion of standard and AI-based computational simulations will obviously incline human collectives to more and more adopt habits and settings that deal with data and problems appropriate to those same tools. This could generate an effect of progressive computational configuration of new social behaviors and environments appropriate to conform themselves to the computational tools and needs, to the detriment of old ones: in few words, will we face eco-cognitive settings which are *computationally tailored*, characterized by the prevalence of locked cognitive strategies or *humanly tailored*, characterized by the unlocked ones?

**GW-Schema** Focusing attention on basic cognitive aspects of abduction, and adopting a logical framework centered on practical agents, (Gabbay & Woods, 2005) contend that abduction (basically seen as a *scant-resource* strategy, which proceeds in absence of knowledge) presents an *ignorance-preserving* (or, better, an *ignorance mitigating*) character. Woods says that of course "[...] it is not at all necessary, or frequent, that the abducer be wholly in the dark, that his ignorance be total. It needs not be the case, and typically is not, that the abducer's choice of a hypothesis is a blind guess, or that nothing positive can be said of it beyond the role it plays in the subjunctive attainment of the abducer's original target (although sometimes this is precisely so)" (Woods, 2013, p. 249). In this perspective, abductive reasoning responds to ignorance-problems. Ignorance-problems are defined as the impossible attainments of certain cognitive targets with the knowledge at one's current disposal. In response to ignorance-problems one can adopt three strategies: one can attain some additional knowledge (subduance), overcoming one's ignorance; one can yield to ignorance (at least for the time being) (surrender); or one can abduce, acquiring a new viewpoint for action, without getting rid of the constitutive ignorance. One of the virtues of the GW-schema also resorts to its capacity to provide the opportunity to illustrate non-explanatory and instrumentalist aspects of abductive cognition. Examples of the non-explanatory features of abduction are for example present in logic and mathematical reasoning. Furthermore, physics often aims at discovering physical dependencies which can be considered explanatorily undetermined. In this case abduction also exhibits an instrumental aspect.

From this perspective the general form of an abductive inference can be symbolically rendered as follows. Let $\alpha$ be a proposition with respect to which you have an ignorance problem. Putting $T$ for the agent's epistemic target with respect to the proposition $\alpha$ at any given time, $K$ for his knowledge-base at that time, $K^*$ for an immediate accessible successor-base of $K$ that lies within the agent's means to produce in a timely way,[5] $R$ as the attainment relation for $T$, $\rightsquigarrow$ as the *subjunctive* conditional relation, $H$ as the agent's hypothesis, $K(H)$ as the revision of $K$ upon the addition of $H$, $C(H)$ denotes the conjecture of $H$ and $H^c$ its activation. The general structure of abduction can be illustrated as follows (GW-schema):

| | |
|---|---|
| 1. $T!\alpha$ | [setting of $T$ as an epistemic target with respect to a proposition $\alpha$] |
| 2. $\neg(R(K, T))$ | [fact] |
| 3. $\neg(R(K^*, T))$ | [fact] |
| 4. $H \notin K$ | [fact] |
| 5. $H \notin K^*$ | [fact] |
| 6. $\neg R(H, T)$ | [fact] |
| 7. $\neg R(K(H), T)$ | [fact] |
| 8. If $H \rightsquigarrow R(K(H), T)$ | [fact] |
| 9. $H$ meets further conditions $S_1, \ldots, S_n$ | [fact] |
| 10. Therefore, $C(H)$ | [sub-conclusion, 1–9] |
| 11. Therefore, $H^c$ | [conclusion, 1–10]. |

It is easy to see that the distinctive epistemic feature of abduction is captured by the schema. It is a given that $H$ is not in the agent's knowledge-set. Nor is it in its immediate successor. Since $H$ is not in $K$, then the revision of $K$ by $H$ is not a knowledge-successor set to $K$. Even so, $H \rightsquigarrow R(K(H), T)$. So we have an ignorance-preservation, as required (cf. Woods, 2013, p. 370).

[5] $K^*$ is an accessible successor of $K$ to the degree that an agent has the know-how to construct it in a timely way; i.e., in ways that are of service in the attainment of targets linked to $K$. For example if I want to know how to spell "accommodate", and have forgotten, then my target can't be hit on the basis of $K$, what I now know. But I might go to my study and consult the dictionary. This is $K^*$. It solves a problem originally linked to $K$.

[*Note*: Basically, line 9. indicates that $H$ has no more plausible or relevant rival constituting a greater degree of subjunctive attainment. Characterizing the $S_i$ is the most difficult problem for abductive cognition, given the fact that in general there are many possible candidate hypotheses. It involves for instance the *consistency* and *minimality* constraints. These constraints correspond to the lines 4 and 5 of the standard AKM schema of abduction.] Finally, in the GW-schema $C(H)$ is read "It is justified (or reasonable) to conjecture that $H$" and $H^c$ is its activation, as the basis for *planned* "actions".

In sum, in the GW-schema $T$ cannot be attained on the basis of $K$. Neither can it be attained on the basis of any successor $K^*$ of $K$ that the agent knows then and there how to construct. $H$ is not in $K$: $H$ is a hypothesis that when reconciled to $K$ produces an updated $K(H)$. $H$ is such that if it were true, then $K(H)$ would attain $T$. The problem is that $H$ is *only hypothesized*, so that the truth is not assured. Accordingly Gabbay and Woods contend that $K(H)$ *presumptively* attains $T$. That is, having hypothesized that $H$, the agent just "presumes" that his target is now attained. Given the fact that presumptive attainment is not attainment, the agent's abduction must be considered as preserving the ignorance that already gave rise to her (or its, in the case for example of a machine) initial ignorance-problem. Accordingly, abduction does not have to be considered the "solution" of an ignorance problem, but rather a response to it, in which the agent reaches presumptive attainment rather than actual attainment. $C(H)$ expresses the conclusion that it follows from the facts of the schema that $H$ is a worthy object of conjecture. It is important to note that in order to solve a problem it is not necessary that an agent actually conjectures a hypothesis, but it is necessary that she states that the hypothesis is *worthy of conjecture*. Briefly, considering $H$ justified to conjecture is not equivalent to considering it justified to accept/activate it and eventually to send $H$ to experimental trial. $H^c$ denotes the *decision* to release $H$ for further work in the domain of enquiry in which the original ignorance-problem arose, that is the activation of $H$ as a positive *cognitive* basis for action.

We have to remember that the process of evaluation and so of activation of the hypothesis, is not abductive, but inductive, as Peirce contended. In the framework of the GW-schema it cannot be said that testability is intrinsic to abduction, such as it is instead maintained in the case of some passages of Peirce's writings. When abduction stops at line 10. (cf. the GW-schema), the agent is not prepared to accept $K(H)$, because of supposed adverse consequences. This activity of testing, which in turn involves degrees of risk proportioned to the strength of the conjecture, is strictly cognitive/epistemic and inductive in itself, an experimental test, and it is an intermediate step to release the abduced hypothesis for inferential work in the domain of enquiry within which the ignorance-problem arose in the first place.

**Heuristic Strategies** We can usefully see selective and creative abduction as performed by the application of "heuristic procedures" (often called "heuristic strategies") that involve all kinds of good and bad inferential moves, beyond the simple mechanical application of rules. It is only by means of these heuristic procedures that the acquisition of new truths is guaranteed. Also Peirce's mature view on creative abduction as a kind of inference seems to stress the strategic component of reasoning. Strategic rules are smart rules, even if they fail in individual cases, and show a propensity for cognitive success. When there is a problem to solve, we usually face several possibilities (hypotheses) and we have to create or select the suitable one. In this situation we can rely on heuristics, that are for example rules of thumb expressed in sentential terms and that help us reach satisfactory choices without considering all the possibilities.

**Hybrid Abducers** The various abductive activities human agents are engaged in drastically rely on external supports. For instance, human agents create mimetic external representations which mirror concepts and problems that are already represented in the brain and need to be enhanced, solved, further complicated, etc. so they can sometimes creatively give rise to new concepts and meanings. In the process of distributing the mind into the material world around them, human agents develop new abductive skills, which would not be exhibited without the continuous interplay with the various external supports. In doing so, human agents can be described as hybrid abducers. Organic agents are spontaneous inducers and abducers but human beings also construct logical and computational (ideal) systems both able to mimic human inductions and abductions and to create new and more "rational" ways of inducing and abducing. These systems are in turn used by human agents: they consequently have to be seen as hybrid reasoners.

**Ignorance, Diagnosticability, Discoverability** A general taxonomy of ignorance realizes the program of the so-called "epistemology of ignorance", which refers to a wide group of problems pertaining human cognitive behavior, also related to discoverability and diagnosticability. Cognitive confinement is a state of being systematically blocked from affordances or even disafforded, and so provided with disfigured affordances due to the way in which the subject's environment (cognitive niche) is set up by the subject-world interactions": we can consider this case a situation of embodied and ecological ignorance. Various cases of "ignorance-making"can be further stressed: (1) ignorance/confinement is—explicitly or implicitly—but intentionally built: for example, filter bubbles in social networks create collectives of people who are ignorant because they are almost totally isolated from facts and opinions that would for example chance their political orientation; (2) ignorance/confinement is the unintentional objective effect of some changes that occur in social equilibria pertaining to specific cognitive niches: the case of recent "epistemic niches" that jeopardize (or tend to jeopardize) discoverability in some areas of science such as biomedicine is an example.

**Ignorant Entities and Overcomputationalization** Our era is characterized by a progressive transformation of ignorant entities not only in cognitive entities but also in computational ones: currently we are facing with an excess of computationalization (overcompulationalization) of the world and we can complain that too many ignorant entities are more and more computationally "domesticated".

**Ignorance-Preserving Abductive Cognition** Abduction is a procedure in which something that lacks classical epistemic virtue is accepted because it has virtue of another kind. For example: let $S$ be the standard that you are not able to meet (e.g., that of mathematical proof). It is possible that there is a lesser epistemic standard $S'$ (e.g., having reason to believe) that you do meet. Focusing attention on this cognitive aspect of abduction we can contend that abduction (basically seen as a scant-resource strategy, which proceeds in absence of knowledge) presents an ignorance preserving (but also an ignorance mitigating) character. Of course it is not at all necessary, or frequent, that the abducer be wholly in the dark, that his ignorance be total. It needs not be the case, and typically is not, that the abducer's choice of a hypothesis is a blind guess, or that nothing positive can be said of it beyond the role it plays in the subjunctive attainment of the abducer's original target (although sometimes this is precisely so). Abductive reasoning is a response to an ignorance-problem: one has an ignorance-problem when one has a cognitive target that cannot be attained on the basis of what one currently knows. Ignorance problems trigger one or other of three responses. In the first case, one overcomes one's ignorance by attaining some additional knowledge (subduance). In the second instance, one yields to one's ignorance (at least for the time being) (surrender). In the third instance, one abduces, and so has a positive and reasoned basis for new action even if in the presence of the constitutive ignorance.

**Induction and Abduction** Since the time of John Stuart Mill, the name given to the most important kinds of conjectural non deductive reasoning has been induction, considered as an aggregate of many methods for discovering generalizations and casual relationships. Consequently induction in its widest sense is an ampliative process of generalization of knowledge. By using induction it is possible to synthesize individual statements into general laws—inductive generalizations—in a defeasible way, but it is also possible to evaluate (confirm or discount) hypotheses (on induction as a form of evaluation cf. the considerations contained in the footnote 2 at p. 134.) In the case of the relationship between abduction, deduction, and induction, illustrated by Peirce and exploited in the so called ST-Model (Select and Test model), induction does not deal with the problem of individuating the ways of "generating" inductive hypotheses but refers to a logic of hypothesis "evaluation". Abduction creates or selects hypotheses; from these hypotheses consequences are derived by deduction that are compared with the available data by induction. In the ST-model induction is used as the process of reducing the uncertainty of established hypotheses by comparing their consequences with observed facts.

**Inference to the Best Explanation** There are two main epistemological/cognitive meanings of the word abduction: (1) abduction that only generates plausible hypotheses (selective or creative)—this is the meaning of abduction accepted in my epistemological ST-model—and (2) abduction considered as an inference to the best explanation, that also evaluates hypotheses by induction. In the latter sense the classical meaning of abduction as an inference to the best explanation (for instance in medicine, to the best diagnosis—selective abduction) is described by the complete abduction-deduction-induction cycle.

**Instinct as Abduction** An example of instinctual (and putatively "unconscious") abduction is given by the case of animal embodied kinesthetic/motor abilities, capable of leading to some appropriate cognitive behavior; Peirce says abduction even takes place when a new born chick picks up the right sort of corn. This is another example, so to say, of spontaneous abduction—analogous to the case of some unconscious/embodied abductive processes in humans. Therefore, instinct vs. inference represents a conflict we can overcome simply by observing that the work of abduction is partly explicable as a biological phenomenon and partly as a more or less "logical" operation related to "plastic" cognitive endowments of all organisms.

**Instrumental Abduction** Abduction exhibits an instrumental and strategic aspect, for instance, when intertwined with the exquisite epistemological problem of the role of unfalsifiable hypotheses in scientific reasoning. In this case, an abductive hypothesis can be highly implausible from the "propositional" point of view and nevertheless it can be adopted for its instrumental virtues, such as in the Newtonian case of action-at-a-distance. Highly implausible hypotheses from the "propositional" point of view can be conjectured because of their high "instrumental" plausibility, where a different role of characteristicness is at stake. We have to note that in some sense all abductions embed instrumental factors. In the general case, one accepts a hypothesis because doing so enables one's target to be attained, notwithstanding that lacks the relevant epistemic virtue of plausibility.

**Locked and Unlocked Strategies** Taking advantage of the concept of abduction, we can contend that the *eco-cognitive openness* is undermined in the case of famous computational programs such as AlphaGo, because they are based on *locked abductive strategies*. Instead, unlocked abductive strategies, which are in tune with what eco-cognitive openness requires, qualify those high-level kinds of abductive creative reasoning that are typical of human-based cognition, locked abductive reasoning strategies are instead much simpler than unlocked ones to be rendered at the computational level.

**Logical Abductive Agents** Logical agents have to be seen as ideal "demonstrative environments", which in turn are suitable tools for the creation of logical/deductive models of abductive reasoning itself. They are the fruit of cognitive externalizations in objective logical systems, endowed with symbolic, abstract and rigorous cognitive features, which are also potentially exploitable at the computational level.

**Manipulative Abduction** Manipulative abduction is a process in which a hypothesis is formed and evaluated resorting to a basically extra-theoretical and extra-sentential behavior that aims at creating communicable accounts of new experiences to integrate them into previously existing systems of experimental and linguistic (theoretical) practices. Manipulative abduction represents a kind of redistribution of the epistemic and cognitive effort to manage objects and information that cannot be immediately represented or found internally. An example of manipulative abduction is the case of the human use of the construction of external diagrams in geometrical reasoning, useful to make observations and "experiments" to transform one cognitive state into another for example to discover new properties and theorems.

**Maximization of Cognition and Abduction** When dealing with abduction in a logical perspective, as an "inferential problem", it is appropriate to adopt the more general concepts of input and output instead of those of premisses and conclusions. From this perspective abductive inferences can be seen as related to logical processes in which input and output fail to hold each other in an expected relation, with the solution involving the modification of inputs, not that of outputs. The chance of finding an abductive solution still appears to depend on the Aristotelian concept of "leading away" (ἀπαγωγή), that is, on the starting of the application of a supplementary logic implementing an appropriate formal inference engine. In this perspective—and given the fact science produces and "maximizes" cognition through a process in which affirming truths implies negating truths—the most important consequence for epistemology is that in scientific abductive reasoning irrelevance and implausibility are not always offensive to reason. For example, we cannot be sure, more broadly, that our guessed hypotheses are plausible (even if we know that looking—in advance—for plausibility is a human good and wise heuristic), indeed an implausible hypothesis can later on result plausible.

**Model-Based Abduction** Inaugurating the new semiotic perspective, Peirce stated that all thinking is in signs, and signs can be icons, indices, or symbols and that all inference is a form of sign activity, where the word sign includes "feeling, image, conception, and other representation" (Peirce, 1866–1913, 5.283). In this light it can be maintained that a considerable part of the abductive processes is model-based. That is, a considerable part of hypothesis creation and selection is model-based: it is occurring in the middle of a relationship between brains and model-based aspects of external objects and tools that have received cognitive and/or epistemological delegations.

**Multimodal Abduction** Multimodal abduction depicts hybrid aspects of abductive reasoning. Abductive inference can be visual as well as verbal, and consequently we have to acknowledge the sentential, model-based, and manipulative nature of abduction. Both evidence and hypotheses can be represented using various sensory modalities. Some basic aspects of this constitutive hybrid nature of multimodal abduction involve words, sights, images, smells, etc. but also kinesthetic experiences and other feelings such as pain, and thus all sensory modalities.

**Naturalized Logic (of Abduction)** A complete revision of mainstream logic is an urgent task to be achieved. This revision will be able to bring logic into a creative rapprochement with cognitive science. This can be achieved by trying to do for logic what over forty years ago Quine and others attempted for epistemology. It is necessary to propose a "naturalization" of the logic of human inference and of the so-called *third-way reasoning*, which refers to those kinds of inference that owe their rightness to the meeting of standards other than deductive validity and inductive strength, certainly overworked and overvalued by logicians, and that instead show the capacity to reach good and reliable results. Third-way reasoning refers to those cases of human reasoning such as abduction, but also to many other kinds of the so-called fallacies, disregarded or simply considered irremediable errors by logic since its ancient and modern beginnings. To understand the naturalization of logic it is very useful to take advantage of the case of a naturalized logic of abduction, which leads us to the emphasis on the importance of the following main aspects: "optimization of eco-cognitive situatedness", "maximization of changeability" of both input and output, and high "information-sensitiveness". Furthermore, a naturalized logic of abduction must acknowledge the importance of keeping record of the "past life" of abductive inferential praxes, contrarily to the fact that traditional demonstrative ideal systems are prototypically characterized by what I call "maximization of memorylessness".

**Niche Construction and Abduction** The theory of cognitive niches describes the environment as a sort of "global market" that provides living creatures with unlimited possibilities. Indeed, not all the possibilities that the environment offers can be exploited by human and non-human animals that act on it. For instance, the environment provides organisms with water to swim in, air to fly in, flat surfaces to walk on, and so on. However, no creatures are fully able to take advantage of all of them. Through the activity of niche construction all organisms try to modify their surroundings in order to better exploit those elements that suit them and eliminate or mitigate the effect of the negative ones. This activity is highly related to the hypothetical virtues provided by abductive cognition. That is, the relevant aspects of the environment are appropriately selected and/or reconstructed so as to turn the local environment—inert from a cognitive point of view—into a cultural mediating structure able to deliver suitable chances for behavior control. Through cognitive niche construction organisms not only influence the nature of their world, but also in part determine the Darwinian selection pressure to which they and their descendants are exposed (and of course the selection pressures to which other species are subjected). It is in this perspective that any abduced hypothesis potentially introduces a change (and an opportunity) in the cognitive processes to advance new perspectives in the coevolution of the organism and the environment.

**Non-Explanatory Abduction** Abduction is not intrinsically explanationist, like for example in the case of its description in terms of inference to the best explanation would suggest. If we maintain that $E$ explains $E'$ *only if* the first implies the second, certainly the reverse does not hold. This means that various cases of abduction are consequentialist but not explanationist. Other cases are neither consequentialist nor explanationist. Non-explanatory modes of abduction are clearly exploited in the "reverse mathematics" where propositions can be taken as axioms because they support the axiomatic proofs of target theorems. Furthermore, often in physics the target is the discovery of physical dependencies which we consider explanatorily undetermined. In this case abduction can exhibit an instrumental aspect. Non-explanatory abductions deal with those situations in which the plausibility of certain hypotheses involves less "propositional" and more "strategical" and instrumental aspects, so that propositional plausibility is lower and strategic plausibility higher.

These cases are far from the clear ones of explanatory abduction that are for example occurring in science and in various kinds of diagnosis.

**Nonmonotonic Reasoning** A logical system is monotonic if the function *Theo* that relates every set of wffs to the set of their theorems holds the following property: for every set of premises $S$ and for every set of premises $S'$, $S \subseteq S'$ implies $Theo(S) \subseteq Theo(S')$. Traditional deductive logics are monotonic: intuitively, adding new premises (axioms) will never invalidate old conclusions. In a nonmonotonic system, when axioms, or premises, increase, their theorems do not. For example, following this deductive nonmonotonic view of abduction, we can stress the fact that in actual abductive medical reasoning, when we increase symptoms and patients' data [premises], we are compelled to abandon previously derived plausible diagnostic hypotheses [theorems], as already—epistemologically—illustrated by the Select and Test model (ST-model). If new information emerges, hypotheses not previously considered can be suggested and a new cycle takes place. In this case the nonmonotonic character of abductive reasoning is clear and arises from the "classical" logical unsoundness of the inference rule: it draws defeasible conclusions from incomplete information.

**Overcomputationalization and Ignorant Entities** See Ignorant Entities and Overcomputationalization.

**Perception as Abduction (and Visual Abduction)** Perception is viewed by Peirce as a fast and uncontrolled knowledge-production procedure. Perception, in this philosophical perspective, is seen as a vehicle for the instantaneous retrieval of knowledge that possibly was previously structured in our mind through more structured inferential processes. By perception, knowledge constructions are so instantly reorganized that they become habitual and diffuse and do not need any further testing. By perception, knowledge constructions are so instantly reorganized that they become habitual and diffuse and do not need any further testing: "[...] a fully accepted, simple, and interesting inference tends to obliterate all recognition of the uninteresting and complex premises from which it was derived" (Peirce, 1866–1913, 7.37). Many visual stimuli—that can be considered the "premises" of the involved abduction—are ambiguous, yet people are adept at imposing order on them. For example we readily form such hypotheses as that an obscurely seen face belongs to a friend of ours, because we can thereby explain what has been observed. This kind of image-based hypothesis formation can be considered as a form of what I have called visual (or iconic) abduction.

**Pregnances as Abductions** In terms of the hypothetical abductive framework we can say that a pregnant stimulus is highly diagnostic and a trigger to initiate abductive cognition, like in the case of the hardwired pregnancy occurring regarding a chicken and its food: the chicken promptly reacts when perceiving it. When a pregnance affects an organism, the abductive reaction can be promptly triggered.

**Selective Abduction** Selective abduction is the process in which a hypothesis is abductively selected from a pre-stored encyclopedia of "abducibles". An example of selective abduction is diagnostic reasoning in medicine: it starts from patient data that is abstracted into clinical features to be explained. Then, selective abduction generates plausible explanatory diagnostic hypotheses.

**Sentential Abduction** Sentential abduction—related to logic and to verbal/symbolic inferences—refers to those situations in which a hypothesis is formed and evaluated relying to the sentential aspects of natural or artificial languages, like in the case of logic. Sentential abduction can be rendered in different ways. For example, in the Peircean syllogistic framework abduction is considered like something propositional and as a type of fallacious reasoning. The sentential models of abduction are powerful but limited, because they do not capture various reasoning tasks which involve creative, model-based, and manipulative aspects.

**ST-Model of Abduction** The so-called Select and Test model (ST-Model) is an epistemological model of medical reasoning, which can be described in terms of the classical notions of abduction, deduction and induction; it describes the different roles played by such basic inference types in developing various kinds of medical reasoning (diagnosis, therapy planning, monitoring). The model is consistent with the Peircean view about the various stages of scientific inquiry in terms of "hypothesis" generation (abduction), deduction (prediction), and induction. The model has been used to implement medical knowledge-based systems of medical reasoning.

**Unlocked and Locked Strategies** See Locked and Unlocked Strategies.

## References

Aliseda, A. (1997). Seeking explanations: Abduction in logic, philosophy of science and artificial intelligence (Doctoral dissertation, Amsterdam: Institute for Logic, Language and Computation).

Aliseda, A. (2006). *Abductive reasoning. Logical investigations into discovery and explanation.* Heidelberg/Berlin: Springer.

Gabbay, D. M., & Woods, J. (2005). *The reach of abduction.* Amsterdam: North-Holland.

Hintikka, J. (2007). *Socratic epistemology. Explorations of knowledge-seeking by questioning.* Cambridge: Cambridge University Press.

Kakas, A., Kowalski, R. A., & Toni, F. (1993). Abductive logic programming. *Journal of Logic and Computation, 2(6)*, 719–770.

Kowalski, R. A. (1979). *Logic for problem solving.* New York: Elsevier.

Kuipers, T. A. F (1999). Abduction aiming at empirical progress of even truth approximation leading to a challenge for computational modelling. *Foundations of Science, 4*, 307–323.

Magnani, L. (2001). *Abduction, reason, and science. Processes of discovery and explanation.* New York: Kluwer Academic/Plenum Publishers.

Magnani, L. (2009). *Abductive cognition. The epistemological and eco-cognitive dimensions of hypothetical reasoning.* Heidelberg/Berlin: Springer.

Meheus, J., Verhoeven, L., Van Dyck, M., & Provijn, D. (2002). Ampliative adaptive logics and the foundation of logic-based approaches to abduction. In L. Magnani, N. J. Nersessian, & C. Pizzi

(Eds.), *Logical and computational aspects of model-based reasoning* (pp. 39–71). Dordrecht: Kluwer Academic Publishers.

Peirce, C. S. (1866–1913). *Collected papers of Charles Sanders Peirce*. Vols. 1–6, C. Hartshorne, P. Weiss (Eds.); Vols. 7–8, A. W. Burks (Ed.) (1931–1958). Cambridge, MA: Harvard University Press.

Woods, J. (2013). *Errors of reasoning. Naturalizing the logic of inference*. London: College Publications.

# Index





L. Magnani, *Discoverability*, Logic, Argumentation & Reasoning 26,
https://doi.org/10.1007/978-3-030-93329-6

## I



## J



## K



## L

The manufacturer's authorised representative in the EU is Springer Nature Customer Service Centre GmbH, Europaplatz 3, 69115 Heidelberg, Germany. If you have any concerns regarding our products, please contact ProductSafety@springernature.com

Printed and bound by CPI Group (UK) Ltd, Croydon, CR0 4YY
30/09/2026
02225513-0014